Disaster Risk Reduction

Methods, Approaches and Practices

Series Editor

Rajib Shaw, Keio University, Shonan Fujisawa, Fujisawa, Japan

Disaster risk reduction is a process that leads to the safety of communities and nations. After the 2005 World Conference on Disaster Reduction, held in Kobe, Japan, the Hyogo Framework for Action (HFA) was adopted as a framework for risk reduction. The academic research and higher education in disaster risk reduction has made, and continues to make, a gradual shift from pure basic research to applied, implementation-oriented research. More emphasis is being given to multi-stakeholder collaboration and multi-disciplinary research. Emerging university networks in Asia, Europe, Africa, and the Americas have urged process-oriented research in the disaster risk reduction field. With this in mind, this new series will promote the output of action research on disaster risk reduction, which will be useful for a wide range of stakeholders including academicians, professionals, practitioners, and students and researchers in related fields. The series will focus on emerging needs in the risk reduction field, starting from climate change adaptation, urban ecosystem, coastal risk reduction, education for sustainable development, community-based practices, risk communication, and human security, among other areas. Through academic review, this series will encourage young researchers and practitioners to analyze field practices and link them to theory and policies with logic, data, and evidence. In this way, the series will emphasize evidence-based risk reduction methods, approaches, and practices.

Takako Izumi · Miwa Abe · Kumiko Fujita ·
Rajib Shaw
Editors

All-Hazards Approach

Towards Resilience Building

Springer

Editors
Takako Izumi
International Research Institute of Disaster Science
Tohoku University
Sendai, Japan

Miwa Abe
Department of Community and Social Studies
Tokai University
Kumamoto, Japan

Kumiko Fujita
International Consortium on Landslides
Kyoto, Japan

Rajib Shaw
Graduate School of Media and Governance
Keio University
Fujisawa, Japan

ISSN 2196-4106 ISSN 2196-4114 (electronic)
Disaster Risk Reduction
ISBN 978-981-97-1859-7 ISBN 978-981-97-1860-3 (eBook)
https://doi.org/10.1007/978-981-97-1860-3

This Springer imprint is published by the registered company Springer Nature Singapore Pte Ltd.
The registered company address is: 152 Beach Road, #21-01/04 Gateway East, Singapore 189721, Singapore

Preface

Eight years have passed since the Sendai Framework for Disaster Risk Reduction was adopted in 2015. In that time, disasters have continued to strike the world—floods in Pakistan, Nepal, India, and Germany; earthquakes in Turkey, Japan, and Afghanistan; wildfires in the United States and Australia; cyclones and typhoons in India and the Philippines, and COVID-19—and the damage and impact never seemed to abate. The experience of COVID-19 made us realize that it is not just important but urgent to consider all hazards in a disaster preparedness/disaster risk reduction plan.

Various organizations and governments have reviewed the progress and challenges in implementing the Sendai Framework to identify what needs to be done by 2030 to achieve the goals. One challenge is that the risk landscape around us is often changing, and our risk management strategy also needs to be more inclusive and comprehensive to address different hazard risks. In addition, disasters often do not occur in isolation. Instead, they have cascading effects, and multiple hazards occur simultaneously, increasing the burden on affected countries. However, there has been little progress in implementing the all-hazards risk assessment described in the Sendai Framework.

First, the all-hazards approach does not mean that equal attention and resources are given to all-hazard risks. It is essential to carry out the assessment by considering all potential risks with the active participation of different stakeholders and experts at different levels. It is crucial to develop a risk reduction strategy that incorporates local experience and voices based on such a thorough assessment. The hazard risks we face are growing and becoming more complex; therefore, managing complexity requires a different approach.

Based on the above context, this book has three main parts: Part I: Overview (consisting of three chapters), Part II: Different Types of Risks (consisting of nine chapters), and Part III: The All-Hazards Approach (AHA) and its Mechanism and System (consisting of six chapters). This book has two specific audiences. The first is students, researchers, and academics in the field of disaster risk reduction, higher education, and risk management studies. The other is practitioners and policymakers who apply collective knowledge to policy and decision-making. The book will contribute to a better understanding of the all-hazards approach and its implementation. We would be grateful if readers find it useful and relevant.

Sendai, Japan Takako Izumi
Kumamoto, Japan Miwa Abe
Kyoto, Japan Kumiko Fujita
Fujisawa, Japan Rajib Shaw

Contents

About the Editors

Takako Izumi has been an associate professor at the International Research Institute of Disaster Science (IRIDeS), Tohoku University, Japan, since 2013. She also serves as a program coordinator of the Multi-Hazards Program under the Association of Pacific Rim Universities (APRU), which comprises 60 universities and academic institutes in the Pacific Rim. She holds a Ph.D. in global environmental study from Kyoto University, Japan. Her research interests include international and regional frameworks and strategies for disaster risk reduction, international humanitarian assistance, and disaster risk reduction (DRR) initiatives at local and community levels. Since May 2015, she has been a member of the Asia-Pacific Scientific and Technical Advisory Group (AP-STAG) to support the efforts of governments and stakeholders in the implementation of the Sendai Framework for DRR.

Miwa Abe became an associate professor at the Faculty of Humanities and Sciences, Tokai University, Japan, in 2022. She received her Ph.D. in global environmental study from Kyoto University, Japan. Her research interests include disaster recovery policy, social–environmental adaptation by resettlement after natural disasters at the local level, and crisis management by national and local administration. Previously, she had 7 year of experience as a firefighter in Kitakyushu City, Japan. After she joined academia, she worked for sustainable community development and community environmental management in Indonesia and other Asian countries under the Institute for Global Environmental Strategies (IGES) Kitakyushu Urban Center as a researcher.

Kumiko Fujita is a researcher at the International Consortium on Landslides. She has more than 20 years of experience in education, research, and development for natural hazard-induced disaster risk reduction, with a Ph.D. in global environmental studies from Kyoto University, Japan. Her expertise is in applying Japanese technologies of mitigating and preventing natural hazard-induced disasters, such as landslides, debris flow, erosion, and flooding, to social implementation in disaster areas in Asian countries. She has coordinated training courses for foreign engineers provided by the Japan International Cooperation Agency (JICA). She also has coordinated science programs offered by the United Nations Educational, Scientific and Cultural Organization (UNESCO) in Indonesia.

Rajib Shaw is a professor in the Graduate School of Media and Governance in Keio University, Japan. He is also a senior fellow of the Institute of Global Environmental Strategies (IGES) Japan and the chairperson of the Sustainable Environment and Ecological Development Society (SEEDS) Asia and the Church World Service (CWS) Japan, two Japanese NGOs. He is also a co-founder of a Delhi (India)-based social entrepreneur start-up, the Resilience Innovation Knowledge Academy (RIKA). Earlier, he was the executive director of the Integrated Research on Disaster Risk (IRDR) and was a professor in Kyoto University. His expertise includes disaster governance, community-based disaster risk management, climate change adaptation, urban risk management, and disaster and environmental education. He is also the coordinating lead author (CLA) for the Asia chapter's 6th Assessment Report of the Intergovernmental Panel on Climate Change (IPCC). He has published 58 books and over 400 academic papers and book chapters.

Part I
Overview

Chapter 1
Introduction and Overview of the All-Hazard Approach

Takako Izumi

Abstract The COVID-19 pandemic posed substantial concerns and questions on our current disaster risk management framework and strategies that mainly focus on natural hazards. We must reconsider our current crisis/emergency/disaster management strategy to tackle more complicated situations, such as—compound, cascading disasters—and different types of hazards including chemical, biological, and technological. This has led to the development of an all-hazard approach (AHA); however, it is not clear how to implement such an approach or what the difference is between a multi-hazard and an all-hazard approach. This chapter aims to clarify these two approaches and discuss how the AHA can be implemented. Finally, the book structure is included to explain the key points of each chapter.

Keywords All-hazards approach · Multi-hazards · Cascading hazards · Compound hazards · Risk assessment

1.1 Introduction

The COVID-19 pandemic posed substantial concerns and questions on our current disaster risk management framework and strategies that mainly focus on natural hazards, especially in Asia and the Pacific (UNDRR 2020). It revealed the international community is unprepared to tackle such a global crisis and questioned the current risk management strategies. In addition to not only the recent experience, but the need to respond to various emergencies such as natural disasters and technological and complex mass-casualty events, has increased during the last decade (Adini et al. 2012).

The Sendai Framework for Disaster Risk Reduction 2015–2030 required a wider scope of environmental, technological, and biological hazards and a more people-centered, all-hazards (AHA), and multi-sectoral approach to disaster risk reduction (DRR) (UN 2015). The AHA has been the principal framework of disaster

T. Izumi (✉)
International Research Institute of Disaster Science (IRIDeS), Tohoku University, Sendai, Japan
e-mail: izumi@irides.tohoku.ac.jp

T. Izumi et al. (eds.), *All-Hazards Approach*, Disaster Risk Reduction,
https://doi.org/10.1007/978-981-97-1860-3_1

planners globally (United Nations 2016; Bodas et al. 2020). This requires transforming current disaster risk management, governance, and practice from focusing on common hazards to looking at all types of hazards.

Nevertheless, an all-hazards approach (AHA) policy has not yet been fully adopted. Capacity-building programs often focus on preparedness for a specific disaster; therefore, legislation, administrative arrangements, and institutional structures are frequently created to respond to that scenario rather than to the common components that characterize different types of emergencies (Adini et al. 2012). Considering different types of hazards, not only natural hazards, and the need to prepare for and respond to them, it is critical to upgrade current risk management strategies and increase the understanding of effective response and mitigation measures. How can we develop and transform the current strategy from a natural-hazards focus to an all-hazard focus? The term “multi-hazard” is often used to cover different types of hazards. However, what is the difference between multi-hazards and all-hazards approaches? This chapter aims to clarify these terminologies and discuss how we should transform the current disaster risk management into all-hazards risk management.

1.2 Multi-hazard Approach

The term “multi-hazard” appeared for the first time in the UN Agenda 21 for sustainable development (Lopez-Saavedra and Marti 2023) as “complete multi-hazard research into risk and vulnerability of human settlements and settlement infrastructure ...” was called to aid pre-disaster planning to human settlement in disaster-prone areas (Lopez-Saavedra and Marti 2023; Kappes et al. 2012).

Later, the Johannesburg Plan (UN 2002) maintained that “an integrated multi-hazard inclusive approach to address vulnerability risk assessment and disaster management including prevention, mitigation, preparedness, response, and recovery is an essential element of a safer world in the twenty-first century.” This Plan was further specified as a risk reduction focus at the second UN World Conference of DRR with the HFA (Lopez-Saavedra and Marti 2023). The framework includes mainstreaming an integrated and multi-hazard approach into development planning and post-disaster or post-conflict phases across relief, rehabilitation, and recovery activities. Furthermore, the Sendai Framework for Disaster Risk Reduction 2015–2030 adopted in 2015 at the 3rd World Conference on DRR’s one of the key guiding principles for the implementation of the framework emphasizes promoting multi-hazard and inclusive risk-informed decision-making for the effective reduction and management of disaster risks (UN 2015).

UNDRR defines multi-hazards as follows: *Multi-hazard means (1) the selection of multiple major hazards that the country faces, and (2) the specific contexts where hazardous events may occur simultaneously, cascadingly, or cumulatively over time, and taking into account the potential interrelated effects. Hazards include (as mentioned in the Sendai Framework for Disaster Risk Reduction*

2015–2030, and listed in alphabetical order) biological, environmental, geological, hydrometeorological, and technological processes and phenomena.

The early multi-hazard concept focused on the interrelationships among multiple natural hazards (Tilloy et al. 2019). With further research, all possible hazards (natural and man-made hazards) have been incorporated into the multi-hazard framework (Zeng et al. 2023). The term "multi-hazard" is used in cases such as (1) different types of natural hazards including earthquakes, floods, typhoons, landslides, wildfire, and droughts, (2) different types of hazards including biological, chemical, industrial, and natural hazards, and (3) cascading and/or compound hazards.

1.2.1 Different Types of Natural Hazards

Terms such as multi-hazards early warning systems (MHEWS) are commonly used. MHEWS can warn of one or more hazards, and such systems' capacity increases the efficiency and consistency of warnings through coordinated and compatible mechanisms and capacities (UN 2016). Most commonly, this case of "multi-hazards" implies different types of natural hazards—earthquakes, floods, landslides, typhoons, volcano eruptions, and forest fires. However, many early warning systems cover only one type of hazard such as floods or cyclones. Given the extreme and unpredictable climate change-related events, the UN has urged countries to invest in multi-hazard early warning systems. Such systems can address several climate-induced disasters that may occur simultaneously (Voice of America 2022). It is essential for warning services to have a multi-hazard perspective (heavy rainfall may trigger not only flooding but also landslides, the warning for which may come from a separate authority) and institutional partnerships among technical agencies (WMO 2018). MHEWS also requires strengthening cooperation and coordination among the agencies involved and empowering local actors (Aguirre-Ayerbe et al. 2020; Jayasekara et al. 2022).

The term "multi-hazard" is also used to refer to vulnerability and risk assessment methods. Most vulnerability and risk assessments concern a single natural hazard; however, risk management takes a multi-hazard approach, considering the possibility of various types of natural hazards (Kappes et al. 2012).

1.2.2 Different Types of Hazards Including Biological, Chemical, Industrial, and Natural

In many cases, the term multi-hazard is used simply to describe different types of hazards beyond natural ones, such as biological, chemical, industrial, and human-induced hazards regardless of whether they occur simultaneously or separately. The

Sendai Framework for DRR also emphasized the need for tackling natural or man-made hazards and related environmental, technological, and biological hazards and risks (UN 2015). This concept has been widely recognized during the COVID-19 pandemic considering the integration of biological hazards and health aspects in DRR strategies. The UNDRR strongly recommends paying attention to multiple hazards and associated disasters such as desert locust invasion, flooding, droughts, and conflicts, which currently pose a serious threat to food security and sustainable development and integrating biological hazards into DRR strategies and mainstream DRR into relevant sectors (UNDRR and African Union 2020).

There is still a gray zone regarding the types of hazards especially whether to include conflicts or terrorist attacks in this hazard definition. In the above document, conflict is based on the country's context where it is one of its major risks. However, the term man-made hazard does not include the occurrence or risk of armed conflicts and other situations of social instability (UNISDR 2018).

1.2.3 Cascading and/or Compound Hazards

Along with the evolution of multi-hazard analysis, other related terms have emerged, understood as multiple simultaneous or non-simultaneous events. When dealing with more than one hazard at a time the terms multi-hazard and compound hazard are often used (Tilloy et al. 2019). Quigley et al. (2020) use the term multi-hazards almost the same as compound events as the occurrence of two or more hazardous events (e.g., an earthquake during COVID-19), indicating concurrent multi-hazard cascades. Compound hazard is a sub-group of the term multi-hazard. The terms multi-hazard and compound hazard have a broad range of interlinked and overlapping definitions. The term compound hazard is sometimes encompassed within multi-hazard distinguished between compound hazard and multi-hazard as follows: "*While compound hazards are characterized as several elements acting together above their respective damage threshold, multi-hazard are characterized as elements of quite different kinds coinciding accidentally or more often following one another with damaging force.*" (Eshrati et al. 2023, p. 79). This explains that multi-hazards often include both compound and cascading hazards.

The term multi-hazard is more frequently employed than compound hazards, most likely because its use is more established (first mentioned in 2002) (van Westen et al. 2002), whereas the term compound was first mentioned in 2012. Also, the terms "trigger" and "cascade" are more often used within the multi-hazard framework (Tilloy et al. 2019) as the early multi-hazard concept is focused on the interrelationship between multiple natural hazards (Zeng et al. 2023). Cascading disasters are of particular interest in multi-hazard studies (Zeng et al. 2023).

As such, "multi-hazard" is used for both compound and cascading hazards, thus it is recommended to clarify the meaning using either compound or cascading hazards if "multi-hazards" imply either of these cases knowing that "multi-hazard" has both meanings.

Multi-hazards are a new concept that is becoming commonplace in risk reduction plans in contrast to the classical approach of considering each hazard and its potential impacts separately (Lopez-Saavedra and Marti 2023). The term is often used; however, shifting from a single-hazard to multiple-hazard approach is not a linear or easily understood process (Valchogiannies et al. 2022). Currently, the research into single hazards is mature. However, when multiple hazards occur simultaneously, with or without interrelations, the results of risk analysis are often inaccurate and incomplete, because multi-hazard risk analysis is not simply the sum of single hazard risk examinations (Lopez-Saada and Marti 2023; Kappes et al. 2012). For instance, even though many hazard assessments refer to multi-hazard concerning the multiple types of hazards to which an area or infrastructure may be exposed, not all studies on multiple hazards have considered all relevant processes of a defined area, rather they are described as more than one hazard approaches, usually without considering the interrelations between hazards or combined impacts (Lopez-Saavedra and Marti 2023; Kappes et al. 2012).

1.3 All-Hazard Approach

After the COVID-19 outbreak, it has become imperative to address the need for an approach and framework that can be adopted for risk management of all types of hazards. The term "all-hazards" is easily understood to include all types and every potential of hazards in its range of consideration as well as all emergency management functions relevant to almost any hazard (Gutierrez and McEntire 2016) regardless of natural, human-induced, technological, environmental, and biological hazards. GA A/77/293 issued in August 2022 also emphasized "*Today's complex and rapidly changing risk landscape demands a shift* ***to incorporate the full scope of hazards*** *outlined in the SFDRR* ***in risk assessment and planning****, including emerging and frontier risks, such as breach of planetary boundaries, rapid and unprepared energy system transitions, rapid technological change, cyber security, and outer space governance, among others*". The Sendai Framework also advocates an all-hazards approach, to which an understanding of the full scope of hazards faced by communities is essential (Norris et al. 2020; Peleg et al. 2021; UNDRR 2020).

Regarding the origin of the term "all-hazard approach (AHA)", it was recognized that the United States needed to integrate its civil defense and emergency management programs in the 1970s. Thus, the term all-hazard was officially coined by the National Governors' Association in 1974 (Gutierrez and McEntire 2016). The Federal Emergency Management Agency (FEMA) adopted AHA and issued the Guide for All-Hazard Emergency Operations Planning in 1996, which recommends considering all types of hazards, disaster phases (mitigation, preparedness, response, and recovery), and actors in emergency management (Gutierrez and McEntire 2016; Waugh 2004; Donahue et al. 2012). More specifically, the guide emphasizes that the list of hazards that must be considered by emergency management should include natural hazards, airplane crashes, power failures, radiological release, terrorism, train derailment, and urban conflagration. In addition, it highlights that it is crucial to maintain an all-hazard capacity without such a concept, and the possibility of excluding or omitting potential, new, and unexpected hazards in the process of hazard identification remains (FEMA 1996).

However, there are objections to this approach mainly stressing that it is dangerous to consolidate different scenarios and arguing that grasping all ends up losing all (Bodas et al. 2020).

There are some keywords to explain the AHA.

1.3.1 Commonalities in Response or Response Preparedness to All Types of Hazards or Comprehensive Disaster Manamgent by AHA

The rationale for such an approach is that there are commonalities among all types of disaster management and that many of the same management strategies can be applied to all emergencies. This also allows for generalized preparedness and planning activities (Peleg et al. 2021; Waugh 2005; Gregory 2015). If the focus is on commonalities and cost-effectiveness, they mainly belong to disaster response and response planning such as warning, evacuation, sheltering, donation management, debris management, and volunteer management (Penta et al. 2020).

In contrast, Fukami (2016) argues that an all-hazard strategy builds those common capacities necessary to prepare for protection against all phases—respond to, recover from, or mitigate natural disaster risks, acts of terrorism, and other man-made disasters—while also building the unique capacities necessary to prepare for, protect against, respond to, recover from or mitigate the risks of specific types of incidents that pose the most substantial risk to the Nation. Waugh (2000) argues that an AHA implies a comprehensive emergency management approach that covers four different phases and functions: mitigation, preparedness, response, and recovery. The underlying principle is that local, state, and federal governments can develop generic disaster programs that can be used under various circumstances (Waugh 2000). For instance, mass evacuation programs can be used to move populations away in case of floods, wildfires, or hazardous material spills. Temporary shelter programs are useful when people are evacuated from severely damaged homes. It is widely recognized that the activities of different disaster management cycles can and should overlap considerably. In the phase of disaster response, for example, it should take a measure to facilitate recovery and encourage preparedness for the next disaster and mitigation to reduce its impact (Waugh 2000). Understanding the connectivity of each phase in the AHA model is crucial.

1.3.2 Strength/Benefit of AHA

The strengths of the AHA include the following:

- The AHA to emergency management is cost-effective, provides an excellent framework for responding to disasters and encourages cooperation among nonprofits, businesses, communities, and various levels of government. It enhances communication and coordination between organizations before and after emergencies, increasing their efficiency and ability to respond (Gregory 2015).
- The AHA attempts to optimize institutional plans and actions across disaster scenarios by trading off specificity for a general reduction in hazard or risk (Donahue et al. 2012).
- The benefits of the AHA are an improvement in the overall level of preparedness for disasters and a reduction in dramatic paradigm shifts resulting from politics. If a single-hazard approach is taken, there will be a visible shift in focus on a particular hazard that causes major damage, accompanied by the redirection of resources. For example, after a terrorist attack in the US, less money was spent on natural hazard mitigation, whereas more time was given to readiness for terrorism involving weapons of mass destruction. This has resulted in reduced preparedness and response capacities to natural hazards (Gutierrez and McEntire 2016).

1.3.3 Weakness of AHA

The weaknesses of AHA include the following:

- The AHA attempts to simultaneously mitigate both natural disasters and terrorism and consolidate scenarios that belong to two separate and distinct categories (Gregory 2015; Peleg et al. 2021).
- Hazards are not evenly distributed across all geographic areas, whereas planning resources are not infinite and decrease rapidly as political jurisdictions become more local (Bodas et al. 2020).
- Under certain circumstances, a general plan may not be as sufficient or effective as a specific plan tailored to a particular type of event. The generalizability of AHA can become a weakness (Gregory 2015).

1.3.4 AHA and AH Assessment

All-hazard assessments are crucial. As has been repeatedly stated, the AHA does not mean that we should cover and prepare for all types of hazards, rather it implies identifying all possible hazards based on an all-hazard risk assessment and analysis. Furthermore, it is crucial to develop a preparedness plan. The plan needs to be reviewed regularly as potential risks will change within a certain timeframe and it is important to ensure that the current plan includes all potential risks.

Some definitions of the AHA highlight this aspect:

- The AHA includes two components: (1) conducting a risk assessment of the jurisdiction to identify potential hazards and risk exposures and their relative probability of occurrence and (2) developing the capacity to deal with multiple hazards through functional planning. The AHA considers all possible hazards as part of risk analysis (Lober 2017).
- The AHA does not mean that organizations must plan for every possible hazard. This implies that organizations should consider all possible hazards as part of their risk analysis. Using a risk-based approach to planning, coupled with functional and prioritized contingency planning, makes the best possible use of limited resources during an emergency (Banda 2018).
- The AHA consists of two elements: (1) identifying potential threats and perils that can adversely impact a community both internally and externally. This is completed through a process known as Hazard Vulnerability Assessment (HVA) and (2) having a standardized model of command and control that will provide the team in an assisted living community with a response framework to promote the successful management of emerging situations, regardless of the root cause of the incident (Szpytek 2023).
- The AHA involves performing a detailed risk assessment of all potential hazards that can affect the organization, and then developing mitigations, planning strategies, and performing testing exercises based on these prioritized hazards (ParadigmSI 2020).
- Four steps to AH planning are as follows: (1) identify hazards, (2) consider your assets, (3) assess impacts, and (4) weigh possible mitigation (Total Security Condition 2019).
- The purpose of the All-Hazards Risk Assessment (AHRA) is to develop a mechanism for the comparative assessment and rating of risk events derived from all hazards to support emergency management planning. The process and methodology primarily focused primarily on the assessment components of the overall risk management paradigm. The risk management steps are as follows: (1) Risk identification, (2) Risk analysis, (3) Risk evaluation, and (4) Risk treatment planning (Verga 2013).
- The AHA should begin with a national risk assessment to develop a risk mapping. The risks identified by the World Economic Forum (WEF) in the global risk report change every year and the number of risks also increases. Each country should conduct a national risk assessment based on the AHA and have a clear command system on who will be responsible for what and when to intervene and respond to emergencies. Therefore, it is necessary to prioritize the risks to be handled (Kishimoto 2015).

The WHO developed the Strategi Toolkit for Assessing Risks (STAR) in 2021 (WHO 2021) based on the principle of AHA "Many emergencies and disaster risk management functions are similar across hazards (e.g., planning, logistics, risk communications), and it is neither efficient nor cost-effective to develop separate, stand-alone capacities or response mechanism for each individual hazard" (WHO 2021).

Step 1: Identify the hazards and describe the scenario most likely to require the activation of a coordinated response.

Step 2: Evaluate the likelihood.

Step 3: Estimate the impact.

Step 4: Determine the risk level.

(Step 5: Finalize recommendations and workshop report.

Step 6: Integrate the recommendations and priority actions into national or subnational action plans for sustainable capacity building.)

Steps 5 and 6 include how to share the risks identified and used in policymaking.

1.4 Key Issues in Aha

The AHA is a systematic approach that includes key principles and steps for implementation such as

(a) The AHA is a risk approach. Therefore, conducting a thorough risk assessment targeting all potential hazards including chemical, technological, and biological hazards and terrorist attacks such as low-frequency and large-scale disasters is essential. Such assessments need to be conducted regularly as risks can change frequently and new risks can emerge. It is ideal to have a general framework for emergency management based on the AHA, and technical guidelines for managing each specific hazard.

(b) Crisis managers and even policymakers must have such a concept not to cause a drastic change in allocating budget and funds toward a specific disaster that has caused tremendous damage recently. This has the risk of reducing the preparedness capacity for other disasters.

(c) The AHA applies to all disaster management cycles including response, recovery, mitigation, and preparedness. It also encourages cooperation among various levels of governments, communities, businesses, and civil societies.

(d) Negative arguments against the AHA include its attempts to consolidate different types of hazards into a single plan. However, having specific plans developed by each agency creates problems, such as conflicts with the plans of other organizations, each crisis is managed by a different manager leaving no clear chain of command and a potentially large number of managers with different objectives. It is more effective to have a consolidated action plan rather than a specific plan in place for every possible scenario. The AHA provides a general framework for emergency management and allows resources to be saved and used more effectively. A general framework document often includes lists of frequent hazards, the roles of responsibilities at that level, a coordination mechanism (incident command system), and emergency management actions (mitigation, preparedness, response, and recovery).

(e) The AHA can be applied by different levels of government.

1.5 Book Structure

This book comprises 18 chapters divided into three major sections. These sections review primary risk management approaches, highlight the characteristics of different types of hazards, examine how hazard risks have been managed at different levels, and discuss the challenges of current risk management and how they can be improved.

Section 1 (Chaps. 1–3) addresses the complexity of future global and cascading risks and the possible challenges and obstacles that countries and individuals may face. It also analyzes the existing disaster management systems and approaches and the steps necessary to strengthen them to prepare for future complicated and cascading hazards. This chapter clarifies the differences between the multi-hazard and all-hazard approach (AHA) and discusses the implementation of the AHA. Chapter 2 highlights how global warming translates into increasing risks in the region, the pathways for resilience, and how investing in these pillars can accelerate the implementation of disaster-related Sustainable Development Goals. Chapter 3 reviews the current risk management approaches to offer a robust pathway for decision-makers to holistically manage the compounding effects of disasters.

Section 2 (Chaps. 4–12) presents case studies of different types of hazard management strategies, including chemical, biological, radiological, and nuclear (CBRN) threats; pandemic hazards; industrial hazards; cybersecurity (natural hazards triggering technological hazards (Natech); planetary health; and systemic risks, and highlights their characteristics. In most cases, different approaches and response mechanisms are applied to manage different types of hazards. A collection of such case studies will help identify common areas of preparedness and response, leading to suitable risk management proposals for the future. Chapter 4 examines CBRN disaster risks, posing key questions on their distinction from natural disasters, the unique risk management that CBRN threats introduce, and how attention to CBRN threats can be integrated into existing all-hazards risk management frameworks. Chapter 5 examines the "all-hazard" applications of Health-EDRM policymaking and program implementation in community resilience-building efforts and discusses current knowledge frontiers and conceptual gaps with the "all-hazard" approaches. Chapter 6 discusses the key Natech cases in Indonesia that severely impacted industries and infrastructure. Chapter 7 focuses on cybersecurity and addresses emerging cyber challenges and the way forward. Chapter 8 summarizes the trends in Natech research worldwide and the future perspective of the Natech risk management approach, introducing catastrophic Natech accidents and reviewing Natech research progress and strategies in each region. Chapter 9 presents a case for defining and addressing climate fragility in a multi-hazard and transboundary context, presenting various short cases to derive messages relevant to climate fragility risk reduction.

Chapter 10 focuses on Afghanistan; examines situations and vulnerabilities specific to food security, climate change, and conflict; and proposes future interventions with AHA perspectives in this context. Chapter 11 discusses the key principles and themes of planetary health, presents key challenges such as climate change, and explores their relevance to disaster management, the humanitarian sector, and the all-hazards approach. Chapter 12 provides an overview of systemic risks, identifies key inflection points in its evolution, and proposes a conceptual framework for enhancing the overall management of systemic risks from a "Whole Systems Approach" using COVID-19 as a case study.

Section 3 (Chaps. 13–18) examines cases of the all-hazards approach at the regional, national, and local levels. Beyond risk management for a single hazard, this section focuses on how the AHA can improve preparedness and response efforts. Chapter 13 examines the issues associated with adopting the AHA by identifying changes in policies and institutions in Japan through in-depth interviews with government officers and researchers in Tokyo, Sendai, and Washington DC. Chapter 14 focuses on compound disasters that have occurred in Japan; examines the existing challenges from the perspectives of the governments, private sector, and local communities; and discusses why AHA is important to overcome these challenges. Chapter 15 focuses on the localization of DRR and Humanitarian Assistance and Disaster Relief (HA/DR), especially from the information-sharing perspective and addresses how localized preparedness and response benefit from broader community expertise and networks. Chapter 16 explains the limitations of scenario-based preparedness from past events and the AHA, based on consequential events to deal with unexpected events, and suggests the effects of prefinance in risk management and a new style of BCP. Chapter 17 introduces the National Disaster Response Force as a case study of the implementation of the AHA in disaster management, showcasing its actual deployment for overseas and domestic missions. Chapter 18 concludes the discussion in this book and highlights the way forward for adopting the AHA to develop a complex risk management framework.

References

Adini B, Goldberg A, Cohen R, Laor D, Bar-Dayan Y (2012) Evidence-based support for the all-hazards approach to emergency preparedness. Israel J Health Policy Res 1:40

Aguirre-Ayerbe I, Merino M, Aye SL, Dissanayake D, Shadiya F, Lopez CM (2020) An evaluation of availability and adequacy of multi-hazard early warning systems in Asian countries: a baseline study. Int J Disaster Risk Reduct 49:101749

Banda F (2018) Mitigating risks with the all-hazards approach to emergency preparedness. file:///C:/Users/Izumi/Downloads/Mitigating_Risks_With_The_All_Hazards_Approach_To_Emergency_Preparedness%20(1).pdf

Bodas M, Kirsch T, Peleg K (2020) Top hazards approach—rethinking the appropriateness of the all-hazards approach in disaster risk management. Int J Disaster Risk Reduct 47:101559

Donahue DA, Cunnion SO, Balaban CD, Sochats K (2012) The all needs approach to emergency response. Homel Secur Affairs 8(1)

Eshrati L, Mahmoudzadeh A, Taghvaeri M (2023) Multi-hazards risk assessment, a new methodology. Int J Health Syst Disaster Manag 3(2)
FEMA (1996) Guide for all-hazard emergency operations planning
Fukami M (2016) An "all-hazards" approach to resilience: benefits and building blocks. https://www.mei.edu/publications/all-hazards-approach-resilience-benefits-and-building-blocks
Gregory PA (2015) Reassessing the effectiveness of all-hazards planning in emergency management. Inq J/Stud Pulse 7(06). https://www.inquiriesjournal.com/a?id=1050
Gutierrez MA, McEntire DA (2016) Emergency management: all-hazards. Encyclopedia of public administration and public policy, 3rd edn. https://doi.org/10.1081/E-EPAP3-120053208
Jayasekara R, Siriwardana C, Amaratunga D, Haigh R (2022) Evaluating the network of stakeholders in multi-hazard early warning systems for multiple hazards amidst biological outbreaks: Sri Lanka as a case in point. Prog Disaster Sci 14:100228
Kappes MS, Patathoma-Kohle M, Keiler M (2012) Assessing physical vulnerability for multi-hazards using an indicator-based methodology. Appl Geogr 32:577–590
Kishimoto A (2015) Integrated risk governance by all hazards approach. ITE Tech Rep 39(15):11–16
Lober JA (2017) All-hazards doesn't mean plan for everything. https://www.iafc.org/membership/iCHIEFS/iCHIEFS-article/all-hazards-doesn-t-mean-plan-for-everything
Lopez-Saavedra M, Marti J (2023) Reviewing the multi-hazard concept. Application to volcanic island. Earth-Sci Rev236:104286
Norris J, Wright N, Fagan L, Mayne L, Murray V (2020) Taking an all-hazards approach to tackling global disaster risk—an important step towards implementing the UN Sendai Framework for disaster risk reduction. EGU2020-22175. https://doi.org/10.5194/egusphere-egu2020-22175
ParadigmSI (2020) Another pandemic?: the importance of the all-hazards approach to planning. https://paradigmsi.com/pandemic-and-all-hazards-approach-to-planning/
Peleg K, Rodas M, Hertelendy AJ, Kirsch TD (2021) The COVID-19 pandemic challenge to the all-hazards approach for disaster planning. Int J Disaster Risk Reduct 55:102103
Penta S, Kendra J, Marlowe V, Gill K (2020) A disaster by any other name?: COVID-19 and support for an all-hazards approach. Risks Hazards Crisis Public Policy 1–26
Quigley MC, Attanayake J, King A, Prideaux F (2020) A multi-hazards earth science perspective on the COVID-19 pandemic: the potential for concurrent and cascading crises. Environ Syst Decis 40:199–215
Szpytek S (2023) "All hazards" emergency management in assisted living communities. Geriatr Nurs 51:448–450
Tilloy A, Malamud BD, Winter H, Joly-Laugel A (2019) A review of quantification methodologies for multi-hazard interrelationships. Earth Sci Rev 196:102881
Total Security Solutions (2019) An all-hazards approach to school safety. https://www.tssbulletproof.com/blog/all-hazards-approach-school-safety#:~:text=The%204%20Steps%20To%20All-Hazards%20Planning%201%20Step,...%204%20Step%204%3A%20Weigh%20Possible%20Mitigations%20
UN (2002) Report of the world summit on sustainable development. Technical report, United Nations
UN (2015) Sendai framework for disaster risk reduction 2015–2030
UN (2016) Report of the open-ended intergovernmental expert working group on indicators and terminology relating to disaster risk reduction
UNDRR (2020) Review of COVID-19 disaster risk governance in Asia-Pacific: towards multi-hazard and multi-sectoral disaster risk reduction. United Nations Office for Disaster Risk Reduction, Regional Office for Asia and the Pacific, Bangkok, Thailand
UNDRR and African Union (2020) Africa COVID-19 BRIEIF: virtual dialogue between national and regional disaster risk reduction focal points
UNISDR (2018) Words into action guidelines: implementation guide for man-made and technological hazards. UN Office for Disaster Risk Reduction
Valchogiannies D, Sfetsos A, Markantonis I, Politi N, Karozis S, Gounaris N (2022) Quantifying the occurrence of multi-hazards due to climate change. Appl Sci 12:1228

Van Westen CJ, Montoya I, Boerboom I, Greiving S (2002) Multi-hazard risk assessment using GIS in urban areas: a case study for the city of Turrialba, Costa-Rica. In: Proceedings regional workshop on best practices in disaster mitigation, 24–26 Sep 2002, Bali, Indonesia, pp 53–72
Verga S (2013) A holistic, cross-government all hazards risk assessment. http://www.ismor.com/cornwallis/cornwallis_2012/SVerga_C17_AHRA.pdf
Voice of America (2022) UN: early warning of impending disasters saves lives. https://www.bignewsnetwork.com/news/272916386/un-early-warning-of-impending-disasters-saves-lives
Waugh WL (2000) Living with hazards dealing with disasters: an introduction to emergency management. Routledge
Waugh WL (2004) The "all-hazards" approach must be continued. J Emerg Manag 2(1)
Waugh WL (2005) Terrorism and the all-hazards model. J Emerg Manag 3(2)
WHO (2021) Strategic toolkit for assessing risks: a comprehensive toolkit for all-hazards health emergency risk assessment. World Health Organization
WMO (2018) Multi-hazard early warning systems: a checklist
Zeng T, Chen G, Reniers G, Hu K (2023) Resilience assessment of chemical industrial areas during Natech-related cascading multi-hazards. J Loss Prev Process Ind 81:104967

Chapter 2
Overview of Disasters, Climate-Related Risks and Resilience Pathways for Asia–Pacific

Sapna Dubey, Madhurima Sarkar-Swaisgood, Maria Bernadet Karina Dewi, and Sanjay Srivastava

Abstract Asia–Pacific region is engulfed with current and impending climate change impacts. With a scientific lens, this chapter highlights some **dynamic risk hotspots and the need to study risks holistically**. Now more than ever, the growing intensity and frequency of natural hazards are leading to cascades affecting livelihoods, health, food security and socio-economic stability of communities. In this context, the spotlight is then directed towards **three key resilience-building policy pathways**. These include recommendations on making risk-informed adaptation investments, developing a grassroots approach for strengthening the health sector response and finally, prioritizing technology and innovation for precise risk analytics and response mechanisms.

Keywords Climate change · Asia Pacific · Cascading risk · Resilience · Adaptation · Health services · Technology

---

S. Dubey (✉) · M. Sarkar-Swaisgood · M. B. K. Dewi · S. Srivastava
Disaster Risk Reduction, UNESCAP, Bangkok, Thailand
e-mail: sapnadubey14@gmail.com

M. Sarkar-Swaisgood
e-mail: sarkar-swaisgood@un.org

M. B. K. Dewi
e-mail: dewi@ehs.unu.edu

S. Srivastava
e-mail: srivastavas@un.org

T. Izumi et al. (eds.), *All-Hazards Approach*, Disaster Risk Reduction,
https://doi.org/10.1007/978-981-97-1860-3_2

2.1 Introduction: Overview of Disasters and Climate Risks

"*The Pakistani people are facing a monsoon on steroids - the relentless impact of epochal levels of rain and flooding*" stated UN Secretary-General António Guterres recently as floods submerged much of the country.

In 2022, more than 1,500 people were killed and 33 million others impacted in one of the country's worst monsoon seasons in over a decade.[1] The Pakistan Meteorological Department reported that the national rainfall was 243% above average, while Balochistan and Sindh provinces registered +590 and +726% precipitation in August 2022.[2]

It was clear that climate change has a role to play in this havoc. Indeed, flooding can be attributed to the extreme monsoon rainfall throughout the summer season (60 days). Along with other drivers, this extreme rainfall was likely caused due to the global warming by 1.2 °C.[3] These floods gave the region a glimpse of an impending climate urgency. Indeed, climate change drives natural hazards in the region causing major economic and non-economic losses. This chapter showcases the disaster and climate change risks in Asia Pacific.

Geophysical, hydrometeorological and socio-economic specificities define the risk landscape of different subregions in Asia–Pacific. For instance, as studied by United Nations Economic and Social Commission for Asia and the Pacific (UNESCAP),[4] arid and semi-arid land has been expanding and aridity intensifying in **South and South-West Asia**, particularly in Afghanistan, the Islamic Republic of Iran, Pakistan and India, as shown in Fig. 2.1.

Given the intense hazard risk in the region, complex cascading impacts on lives, livelihoods, economies and people come into play. For instance, in 2021 the Asia region witnessed more than 100 natural hazards, the majority of which were flood events. These caused nearly 4000 fatalities. Overall, the hazard touched 48.3 million people, causing USD 35.6 billion economic damage.[5] At the country level, **Pakistan** is home to some arid zones like Balochistan with a high dependency on rainfall. Rainfall deficits between April and November 2021 resulted in drought-like situation in the region. This turned more challenging with inflation in food and fuel prices, livestock diseases and the impacts of the COVID-19 pandemic. Between March and June 2021, approximately 571,000 people faced a food insecurity crisis stage and around 185,000 people faced an emergency stage.[6] Similarly, **tropical Cyclone Ana** wreaked havoc in various parts of **Fiji**, with strong winds and heavy

[1] United Nations ESCAP (2022b).

[2] Ibid.

[3] World Weather Attribution (2022).

[4] Ibid.

[5] World Meteorological Organization (2022a).

[6] World Meteorological Organization (2022a).

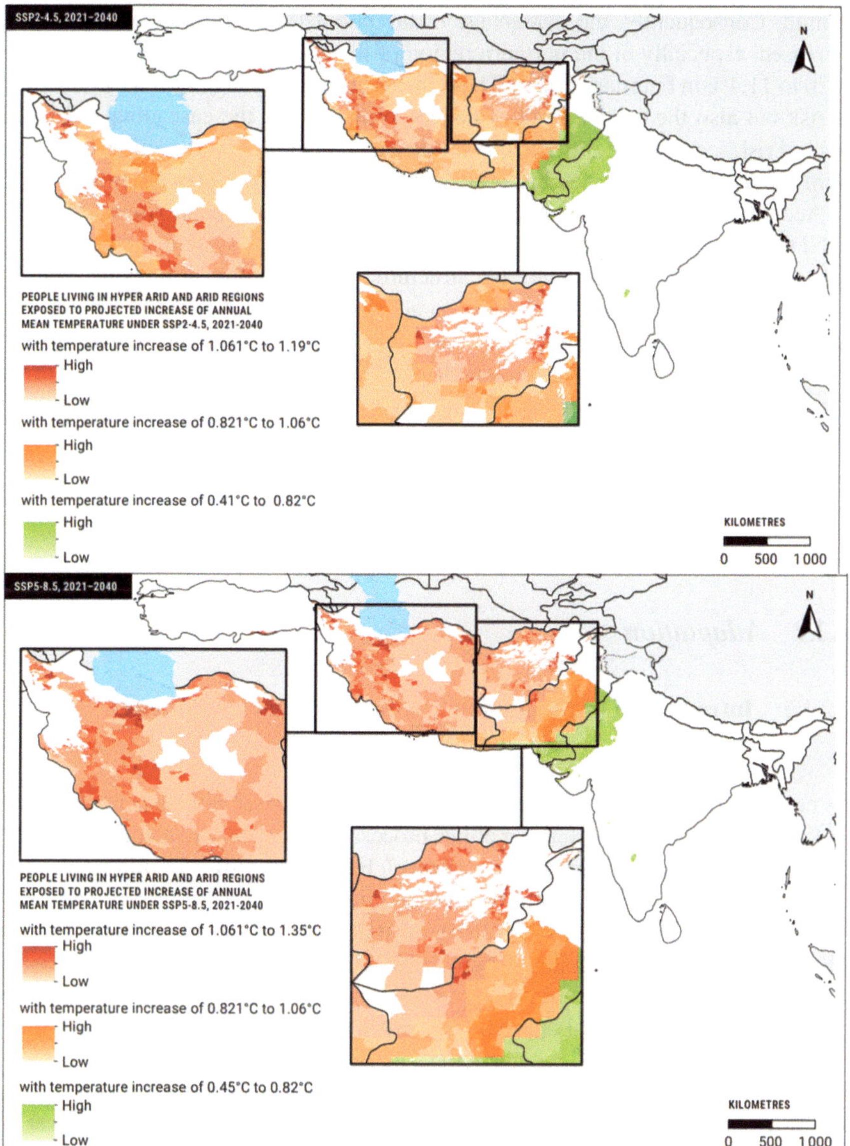

Sources: ESCAP calculations, based on IPCC WGI Interactive Atlas - Coupled Model Intercomparison Project Phase 6 (CMIP6), 2021; The Global Aridity Index, 2019; UN-WPP Adjusted Population Count 2020, v4.11; and UN Geospatial.

Notes: 1. Projected increase 2021-2040 in Annual Mean Temperature Change under SSP2-4.5 ranges from 00.41°C to 1.19°C, and from 0.45°C to 1.35°C under SSP5-8.5.
2. The reference period for the projected change 2021-2040 in Annual Mean Temperature Change is 1995 - 2014.

Disclaimer: The boundaries and names shown and the designations used on this map do not imply official endorsement or acceptance by the United Nations. Dotted line represents approximately the Line of Control in Jammu and Kashmir agreed upon by India and Pakistan. The final status of Jammu and Kashmir has not yet been agreed upon by the parties.

Fig. 2.1 Population exposed to hyper-arid and arid regions with additional exposure to projected increase of annual mean temperature under two climate change scenarios

rainfall. Consequently, the percentage of the population suffering food insecurity increased, especially in the northern region of the country from 4.2% in December 2020 to 11.4% in February 2021.[7] These instances not only highlight the complexity of risk but also the need for policymakers to understand the cascading and multi-layered risk scenarios in order to protect the ecosystem. The next section transitions from discussing the complex hazard risks to methods to develop resilience.

According to the United Nations Framework Convention on Climate Change (UNFCCC), **resilience** is the ability of a social or ecological system to absorb disturbances while retaining the same basic structure and ways of functioning, the capacity for self-organization and the capacity to adapt to stress and change.[8] The chapter explores **three policy pathways** for effective resilience building, **first**, making risk-informed adaptation investments, **second**, developing a transdisciplinary and grass-roots approach for strengthening public health service delivery and **third**, prioritizing technology and innovations for better precision in risk analytics.

2.2 Resilience Pathways

2.2.1 Adaptation

2.2.1.1 Introduction to Adaptation as a Key Solution for Climate Resilience

As noted in the UNFCCC, adaptation refers to human-driven adjustments in ecological, social or economic systems or policy processes, in response to actual or expected climate stimuli and their effects or impacts.[9] **Hence, investing in adaptation is a key policy driver to tackle climate change impact**.

However, the Sharm El-Sheikh Implementation Plan, part of the global Conference of Parties (COP27) held in Egypt in November 2022, highlighted the existing gap between current and required adaptation measures in place. It urges all countries to adopt a transformational approach to enhancing adaptive capacity and building resilience.[10] Despite the magnitude of risk, the UNEP *Adaptation Gap Report 2022*[11] shows that international adaptation finance flows to developing countries are five to ten times below estimated needs and will need over US$300 billion per year by 2030. This brings out a strong need to set up mitigation and adaptation measures in

---

[7] World Meteorological Organization (2022b).

[8] United Nations Framework Convention on Climate Change (UNFCCC).

[9] United Nations Framework Convention on Climate Change (UNFCCC).

[10] United Nations Framework Convention on Climate Change (2022).

[11] United Nations Environment Programme (2022).

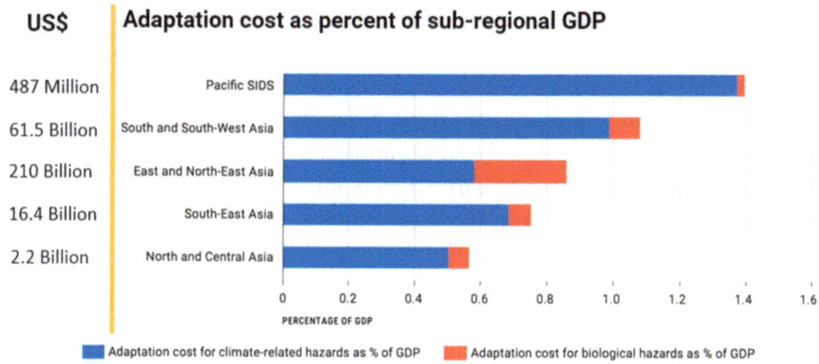

Fig. 2.2 Adaptation costs by ESCAP subregions, in US$, as per cent of subregional GDP. *Source* UNESCAP (2021), resilience in a riskier world: Asia Pacific disaster report 2021

the field of planning, finance and implementation that enhance the co-benefits in all sectors such as agriculture, forestry, ecosystems, water and energy. To progress on bridging the adaptation gap, the following section presents UNESCAP analysis on **cost of adaptation and the key adaptation priority sectors** for investment in order to build resilience.

2.2.1.2 Adaptation Costs and Priorities in Asia–Pacific

In the *Asia Pacific Disaster Report, 2021*, UNESCAP estimates the adaptation cost for all countries and subregions based on annual adaptation cost for natural and biological hazards in the worst climate change scenario. An overview of these costs is presented in Fig. 2.2.[12] For instance, at the country level, the **highest annual adaptation costs** are estimated for China at US$188.8 billion followed by India at US$46.3 billion, Japan at US$26.5 billion, Indonesia at US$8.8 billion and the Republic of Korea at US$8.6 billion. However, the picture alters when these costs are estimated as a percentage of the country's GDP. **As per cent of GDP**, the highest figures are noted for Vanuatu at 9.6%, Tonga at 8.6%, Micronesia (Federation of) at 4.7%, Fiji at 3.5% and Solomon Islands at 3.1%.[13]

[12] UNESCAP (2021).

[13] Ibid.

These investments must be risk-informed and directed into implementing and operationalizing adaptation solutions that yield high investment benefits. UNESCAP analysis presented in Fig. 2.3 gives an overview of the top adaptation priorities for each sub-region, based on the 5 action areas with the highest investment benefits established by the Global Commission on Adaptation.[14] This analysis emphasizes that adaptation-related investments must be informed by the local risk landscape of each region. Hence, these priorities vary with the changing climate and hazard landscapes. For example, while strengthening early warning systems and making new infrastructure resilient are the top priorities in the high precipitation and cyclone-prone South and South West Asia, focusing on improving dryland agriculture crop production and increasing water resource management are the highest priority in the drought-prone North and Central Asia.[15]

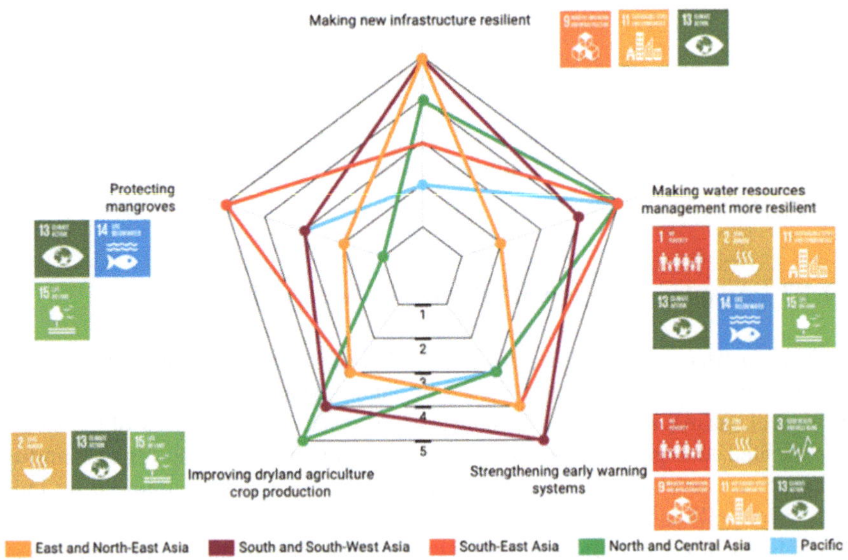

Fig. 2.3 Adaptation priorities for Asia–Pacific subregions with co-benefits for associated United Nations Agenda 2030 Sustainable Development Goals (SDGs).[16,17] *Source* Adapted from UNESCAP (2021), resilience in a riskier world: Asia Pacific disaster report 2021

[14] Global Commission on Adaptation (2019).
[15] UNESCAP (2021).
[16] Ibid.
[17] United Nations, Department of Economic and Social Affairs.

Box 1: Operationalizing UN Secretary General's Call for 'Early Warning for All' That Leaves No-One Behind

In March 2022, UN Secretary-General Antonio Guterres announced that the UN will spearhead the target of achieving early warning for all within the next five years. The Executive Action Plan launched at COP27 in Egypt calls for investments of **USD 3.1 billion** over the next five years to develop fully functional and integrated **Multi-Hazard Early Warning Systems (MHEWS)**.

The Global Commission on Adaptation Report, 2020 specifically highlights that investing in early warning systems provides a high benefit-to-cost ratio of 9:1 globally.[a] Hence, early warning systems are a key climate adaptation measure that saves lives and yields a high return on investment. However, there remain significant gaps in their adequate integration, including in the Asia Pacific region and especially in the Small Island Developing States (SIDS).

Among the Pacific SIDS, only Samoa, Tonga, Tuvalu and Solomon Islands report having MHEWS in place.[b] Further, the SDG score of the related sub-indicator on 'health capacity and emergency preparedness' brings forth the gap in the health sector in the Pacific SIDS.[c] Indeed, **this reiterates the need for investment and policy actions to operationalize effective MHEWS that include biological hazards, in the region**.

Box figure: International Health Regulation Capacity Indicator, 2020

(SDG Target 3.d: Health capacity and emergency preparedness)

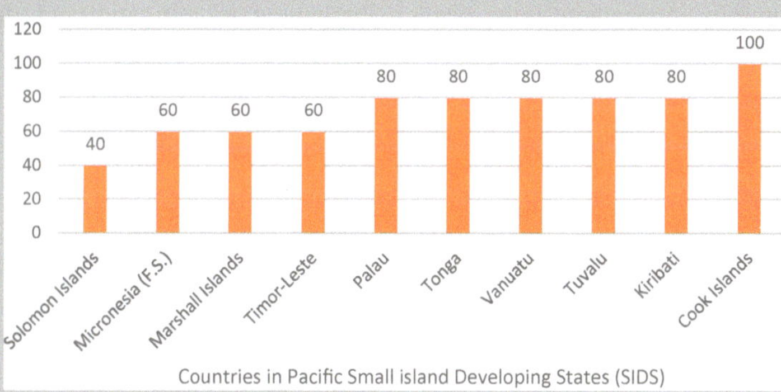

Sources

[a]Global Commission on Adaptation. Adapt Now: A Global Call for Leadership on Climate Resilience. (2020) Available at: https://gca.org/about-us/the-global-commission-on-adaptation/

[b]World Meteorological Organization (WMO) (2022), Early Warnings for All: Executive Action Plan 2023–2027, Available at: https://library.wmo.int/index.php?lvl=notice_display&id=22154#.Y5hRZHZByUl

[c]UNESCAP, SDG Gateway, Asia Pacific, Available at: https://dataexplorer.unescap.org/

2.2.2 *Lateral Public Health Systems*

One of the biggest lessons of the pandemic is that in the pursuit to deliver the best health services, the hospitals and the health sector cannot operate independently. Further, compounding risks and hazards worsens the capacity to respond. Hence, in the study on lateral public health, Professor Jan C. Semenza emphasises on a multi-stakeholder approach that involves the community, one of the key stakeholders to develop resilience and address hazard, exposure and vulnerability.[18] For example, a lateral public health approach to climate-sensitive infectious disease outbreaks due to changing weather patterns and related hazards includes operationalizing a community-based surveillance for early detection of an outbreak at community level. Another example of such a method to tackle water-borne outbreaks arising from drought or water contamination due to environmental degradation is developing a community-based water harvesting and purification system through low-cost chlorination interventions.[19] Overall, this approach relies on multi-sectoral collaboration and community engagement as the means to implement risk mitigation and adaptation methods to tackle hazards.

2.2.3 *Frontier Technologies and Digital Solutions*

Frontier technologies and digital innovations are the backbone of effective risk analytics like forecasting and disaster early warning, surveillance and impact assessment, as well as for managing pandemics, like COVID-19. These technologies can be transformative in nature. Some of the top approaches and related tools of operationalization are **first**, by extending the reach and capabilities of unmanned vehicles, robotics, remote and in situ sensing; **second**, by changing how things are made and acquired, through additive manufacturing and innovative materials and **third**, by strengthening connections between people, things and information, for example in cloud computing, 5G Mobile Technology, wireless mesh networks, mobile messaging, the Internet of Things and blockchain[20] (Fig. 2.4).

[18] Semenza and Shlomit (2021), Semenza (2021).

[19] Ibid.

[20] UNESCAP (2022a).

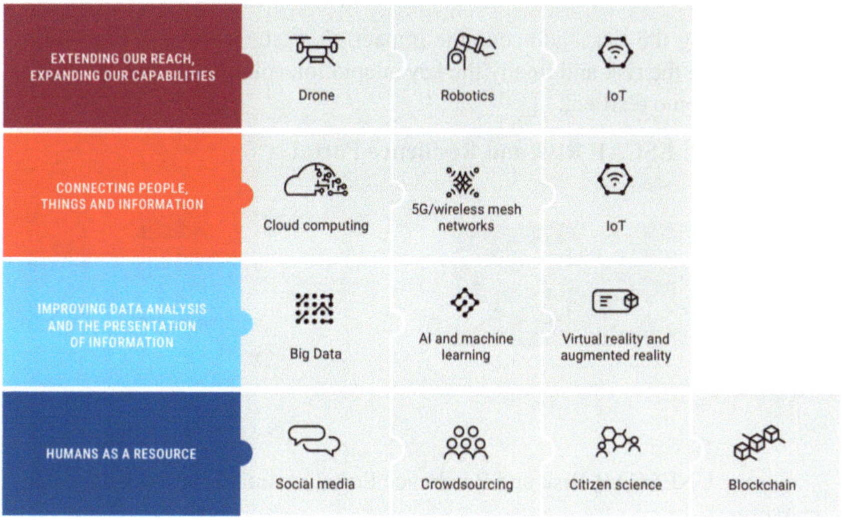

Fig. 2.4 Frontier technologies for disaster risk reduction. *Source* UNESCAP (2022a), Asia-Pacific riskscape @1.5 °C: subregional pathways for adaptation and resilience[21]

Box 2: ESCAP Risk and Resilience Portal: An Interactive Data Portal

The ESCAP Risk and Resilience Portal is a digital platform serving as a one stop shop for policymakers to understand the regional, sub-regional and national 'riskscape' of hazard, climate risks and utilize adaptation measures to build resilient economies and safeguard populations. It converts the vast array of publicly available geospatial, statistical and remote sensing information on hazards, climate change, socio-economic data into usable and inter-operable data analytics for risk informed decision-making. It further provides targeted adaptation solutions for countries based on the risk profile. These analytics can facilitate monitoring of climate and disaster related SDG targets, especially SDG1, SDG2, SDG3, SDG9, SDG11, SDG13 and SDG15 of the United Nations Agenda 2030.

Statistical applications of global spatial datasets on natural and biological hazards, health-related indicator and socio-economic vulnerability indices in QGis are used to identify pockets of **complex and cascading risk hotspots** at the regional, subregional and national level. **Economic impact** is estimated based on probabilistic methods designed to calculate average annual loss in the current term and under two climate scenarios—moderate and worst-case scenario. Finally, targeted **adaptation solutions** are derived based on indexed indicators used to estimate the cost and priorities of five key adaptation measures. Overall, the Portal takes the user on a journey from first,

[21] Ibid.

understanding the risk and economic impact, then, the estimated adaptation cost to tackle the risk and finally the key adaptation solutions to minimize the risk and become resilient.

Box figure 2: ESCAP Risk and Resilience Portal

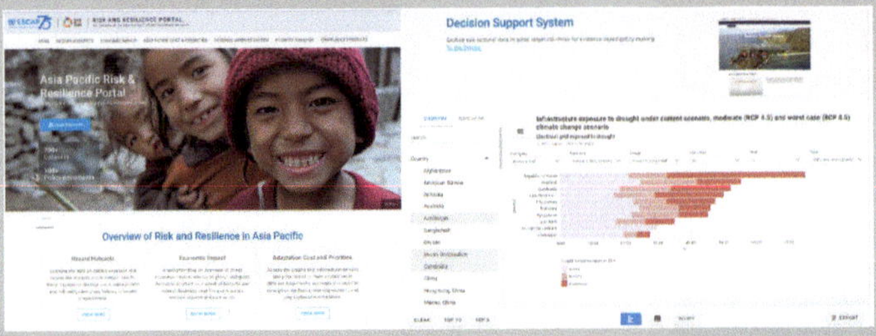

Source UNESCAP, Risk and Resilience Portal, Available at: https://rrp.unescap.org/

2.3 Conclusion

Overall, as explored in the chapter, now more than ever, preparedness for complex and cascading risk scenarios and the way forward must be rooted in an 'all hazards approach'. For instance, early warning systems must be enabled to bring forth multi-hazard risk scenarios. Our health infrastructure must be prepared to tackle multiple simultaneous stresses as witnessed during COVID-19 pandemic. It has been established that shared vulnerabilities and risks cut across borders and boundaries. Hence, these resilience pathways should also be rooted in **transboundary and regional cooperation to attain the best results**. Operationalizing these approaches is then set to enable robust, sustainable, resilient communities and ecosystems.

References

Global Commission on Adaptation (2019) Adapt now: a global call for leadership on climate resilience, 13 September 2019. https://gca.org/reports/adapt-now-a-global-call-for-leadership-on-climate-resilience/

Intergovernmental Panel on Climate Change: The Physical Science Basis (2021) Summary for policymakers. https://www.ipcc.ch/report/ar6/wg1/chapter/summary-for-policymakers/

Semenza JC (2021) Lateral public health: advancing systemic resilience to climate change. https://www.ncbi.nlm.nih.gov/pmc/articles/PMC8495299/

Semenza JC, Shlomit P (2021) Climate change and infectious disease in Europe: impact, projection and adaptation. https://www.thelancet.com/journals/lanepe/article/PIIS2666-7762(21)00216-7/fulltext

United Nations Economic and Social Commission for Asia and the Pacific (UNESCAP) (2017) Asia-Pacific disaster report. https://www.unescap.org/publications/asia-pacific-disaster-report-2017-leave-no-one-behind

UNESCAP (2021) Resilience in a riskier world: Asia Pacific disaster report 2021. https://www.unescap.org/kp/2021/asia-pacific-disaster-report-2021

UNESCAP (2022a) Asia-Pacific riskscape @1.5 °C: subregional pathways for adaptation and resilience, Asia Pacific disaster report 2022 for escap subregions, summary for policymakers. https://www.unescap.org/kp/2022/asia-pacific-disaster-report-2022-escap-subregions-summary-policymakers#:~:text=The%20Asia%20Pacific%20Disaster%20Report,warming%20winds%20with%20intensifying%20tropical

UNESCAP (2022b) Managing riskier Asian monsoon is key to region's resilience. https://www.unescap.org/blog/managing-riskier-asian-monsoon-key-regions-resilience

United Nations, Department of Economic and Social Affairs. https://sdgs.un.org/goals

United Nations Environment Programme (2022) Too little, too slow: climate adaptation failure puts world at risk, adaptation gap report 2022. https://www.unep.org/resources/adaptation-gap-report-2022

United Nations Framework Convention on Climate Change (2022) Sharm el-Sheikh implementation plan. https://unfccc.int/documents/624444

World Meteorological Organization (2022a) State of the climate in Asia 2021. https://library.wmo.int/records/item/58229-state-of-the-climate-in-asia-2021

World Meteorological Organization (2022b) State of the climate in South-West Pacific 2021. https://library.wmo.int/records/item/58225-state-of-the-climate-in-south-west-pacific-2021#.Y8lGrezMLjC

World Weather Attribution (2022) Climate change likely increased extreme monsoon rainfall, flooding highly vulnerable communities in Pakistan. Available at: Climate change likely increased extreme monsoon rainfall, flooding highly vulnerable communities in Pakistan—World Weather Attribution

Chapter 3
History of Risk Management Approach

Tanaya Sarmah, Krishnakali Ghosh, Ranit Chatterjee, and Rajib Shaw

Abstract During the last decade, the studies related to disaster risk management approaches have had a paradigm shift from looking at a single hazard risk at a time to an all-hazards approach wherein multiple complex hazard risks are dealt with comprehensively. This shift is more profoundly mentioned in Priority 1 of the post-2015 Sendai Framework for Disaster Risk Reduction (SFDRR), which is a global document and emphasizes considering an all-hazards approach to assess disaster risks and thereby strengthen capacity. With the growing changes in climatic conditions and human-related activities, the frequency and severity of compounding, cascading, and systemic hazard risks have also significantly increased. The ongoing COVID-19 pandemic since 2020 coupled with various other disasters such as floods, cyclones, heat waves and locust attacks in the global scenario is a well-set example of the terrifying consequences that such complex hazard risks may cause. This chapter systematically reviews the evolution of disaster risk management approaches over time—from dealing with a single hazard risk at a time to an all-hazards approach. This gives an overview of the multiple threats the society and infrastructures are exposed to at a time and their readiness to withstand the same. The chapter also discusses the alternative top-hazards approach which prioritizes the hazard risks according to the local scenario and then sequentially deals with the top-ranking hazard risks for achieving better levels of preparedness. Accordingly, this puts forth the fact that different hazard risks should be managed with a common adaptable plan for mitigation, preparedness, response, and recovery, and also demands an upgradation in the elements of disaster risk management in each of these four stages. A review of the risk management approaches to date shall therefore offer a robust pathway to the decision-makers for managing such compounding effects of disasters holistically and contribute to resilience building of the society and infrastructures therein.

T. Sarmah
Cranfield University, Bedford, UK

K. Ghosh · R. Chatterjee (✉)
Resilience Innovation Knowledge Academy (RIKA), New Delhi, India
e-mail: ranit13@gmail.com

R. Shaw
Keio University, Tokyo, Japan

T. Izumi et al. (eds.), *All-Hazards Approach*, Disaster Risk Reduction,
https://doi.org/10.1007/978-981-97-1860-3_3

Keywords All-hazards · Compound risk · Risk management · Resilience building · Top-hazards

3.1 Introduction

As the world becomes ever more populated and human settlements continue to expand at alarming rates, the impacts of natural hazards have increased sharply over the last half-century. Based on the EM-DAT database, from 1900 to the present, more than 22,000 mass disasters have occurred around the world (Rajabi et al. 2022). Indeed, globalization and the concentration of population around large cities increases exposure to climate and other natural-hazard risks. The nature of these disasters is such that they are not only increasing in frequency and severity, but they are also becoming one of the major challenges in the overall development of a society (IPCC 2022). Globally, natural hazards have caused an estimated 931,000 direct fatalities and over EUR 3.87 trillion in economic losses over the last 20 years (CATDAT 2021) and have affected more than 200 million people per year on average (CRED 2021). Since the 1970s, the idea that disasters are also a human construct has become accepted and is now well embedded in the literature (Kelman et al. 2016). As a response to this, mankind has always attempted to adopt measures to deal with such adverse situations. These actions have resulted in the concept and approaches of disaster risk management (DRM) evolving throughout history. Since the concept of DRM has constantly evolved, it seems pertinent to understand the reasons behind this change and the indicators or events leading to these changes over time. Knowing this will also help to understand the different challenges society has been facing over the years. Therefore, this study tries to identify these paradigm shifts that can help society to adopt effective, logical, scientific and evidence-based approaches for resilience building.

Recent decades have seen a move from managing hazards to managing risks, as documented in several works. The vast majority of that research has focused on single hazards. Internationally, there is an ongoing call for more attention to multi-hazard risk. It should be noted that there are many different terminologies used in this rapidly evolving field. Table 3.1 shows several key definitions of the terms multi-hazard, multi-hazard risk and multi-risk. The definition of disaster risk has evolved with time as new understandings and experiences have emerged. Initially, disaster risk was primarily viewed as a natural phenomenon that resulted in the loss of life and property damage. However, with advances in research and the recognition of social, economic and environmental factors, disaster risk is now understood to be a complex and multi-dimensional concept. The risk definition has changed over the years which has been instrumental in developing approaches for risk management.

In the 1980s, researchers began to focus more on the social and economic factors that influence disaster risk, such as poverty, inequality and inadequate infrastructure (Cutter et al. 2003). In the 1990s, the concept of vulnerability was emphasized to understand the social and economic conditions that make people and communities

Table 3.1 Key definitions related to multi-hazard risk (Ward et al. 2022)

Term	Definition	References
Multi-hazard	The selection of multiple major hazards that the country faces, and the specific contexts where hazardous events may occur simultaneously, cascading or cumulatively over time, and taking into account the potential interrelated effects	UNDRR (2022)
Multi-hazard risk	Risk generated from multiple hazards and the interrelationships between these hazards (but not considering interrelationships on the vulnerability level)	Zschau (2017)
Multi-risk	Risk generated from multiple hazards and the interrelationships between these hazards (and considering interrelationships on the vulnerability level)	

more susceptible to the negative impacts of disasters (Wisner et al. 2004). In the early 2000s, the concept of resilience emerged as a way to understand how communities and systems can adapt and recover from disasters (Adger 2000). In recent years, there has been a growing recognition of the need to address the underlying causes of disaster risk, such as climate change, environmental degradation and unsustainable development practices (IPCC 2018). In addition, there is an increasing emphasis on the importance of local knowledge and community participation in disaster risk reduction (DRR) efforts (UNDRR 2015). The COVID-19 pandemic has also highlighted the importance of considering public health and disease outbreaks as a part of disaster risk (WHO 2020). These changes in the definition of disaster risk have led to a more comprehensive understanding of the factors that contribute to disasters and the need for a holistic and integrated approach to DRM.

3.2 Evolution of DRM Approaches

Evolution in the DRM approaches started with the emergence of civil defense during the twentieth century (Rajabi et al. 2022). Gradually several civil defense units were assigned to manage disasters. Following this, legal frameworks were created to support these units thus forming the modern management of disasters. Initially, DRM was focused on responses but with the evolution in its concepts and understanding, DRR includes prevention and mitigation, whilst disaster management includes response and recovery measures.

On December 11, 1987, the United Nations announced the 1990s as 'The International Decade for Natural Disaster Reduction (IDNDR)', to raise awareness of the importance of reducing disaster risk. IDNDR promoted efforts to build capacity and reduce losses to resources after a disaster and especially in developing countries. The evolution of DRM approaches can be said to depend on three important events—the Yokohama strategy, Hyogo Framework for Action (HFA) and Sendai Framework for Disaster Risk Reduction (SFDRR).

The interaction of different hazards can lead to an impact that is greater than the sum of the single hazard effects. The Yokohama strategy and its plan of action were formulated to create a more secure world, as it was recognized that humanity needs world-class risk management. While Hewitt and Burton (1971) noted a half-century ago the need to shift natural hazard research from a single-hazard approach towards a systematic multi-hazard approach, it surprisingly came into effect only in the early 1990s. Agenda 21 for sustainable development (UNCED 1992) called for a 'complete multi-hazard research' approach to human settlement planning and disaster risk. This notion took a centre stage in the HFA and was followed by the SFDRR. SFDRR calls to 'promote investments in innovation and technology development in long-term, multi-hazard and solution-driven research in DRM'.

The HFA (2005–2015) influenced the global community to adopt a more comprehensive and holistic approach to mitigate disaster risk. It also encouraged countries to pursue three strategic goals, which included (1) Integrating DRR programmes into all sustainable development policies and programmes, (2) Developing and strengthening institutions, mechanisms and capacities to create resilience against hazards and (3) The systematic implementation of risk reduction approaches to emergency preparedness, response and recovery activities (UNDRR 2007).

SFDRR is aimed at the local, national, regional and global levels in order to make people, communities and countries resilient (UNDRR 2015). SFDRR is an authoritative international document for action by 2030 that emphasizes an all-hazards approach to access disaster risks and thereby strengthen capacity. It describes the four stages of DRM—mitigation, preparedness, response and recovery. It envisages that if sufficient capacity is created for the four stages of DRM, then hazard, exposure and vulnerability management can reduce disasters. SFDRR emphasizes that any policy and practice should be based on understanding the concept of risk. In this regard, SFDRR emphasizes designing risk management approaches tailored to each community in accordance with its conditions and adjusted to the characteristics of that specific community.

Additionally, recent years have seen the development of several networks focusing on various aspects of multi-hazard and multi-risk assessment and management. A glaring example is UNDRR's Global Risk Assessment Framework (GRAF) which has a dedicated Working Group to Foster Systems Thinking for multi-hazard risk assessment. Similarly various research programmes (Future Earth, IRDR (Integrated Research on Disaster Risk), WCRP (World Climate Research Programme) and WWRP (World Weather Research Programme)) have established interdisciplinary network to assess and address systemic risk. In similar veins, the climate community have focused on compound climate extremes to address early warning and impacts (Ward et al. 2022).

3.2.1 Single Hazard Versus Multi-hazard Approach

The community of risk professionals have made great advancements in the field of identifying hazards and associated risks, but the progression from a single-hazard approach to a multi-hazard approach is still lacking in the domain. This has proven to be challenging for risk professionals and decision-makers to address real-world problems such as (1) Multiple hazards can lead to a chain of cascading hazards and amplify the associated risks. For instance, a landslide could be triggered by an earthquake; high temperature could increase the probability of wildfires; rainfall and storm surge together could lead to compound flooding or a region could face a combination of such hazards, with varied exposure and/or vulnerability, (2) DRM strategies adopted to reduce the risk from one hazard may pose to be a threat in corresponding to another hazard. For instance, wood or timber frame buildings perform well in earthquakes but could sustain high damages or prove to be fatal during a flood or fire situation and (3) These interconnected impacts due to multiple hazards have consequences across varied sectors and regions. For example, trade-offs may be required while meeting the collective demand and associated supply for sustainable water, energy and food whilst reducing the hazard risks (Ward et al. 2022). Figure 3.1 illustrates the progression from a single-hazard approach to a multi-hazard approach.

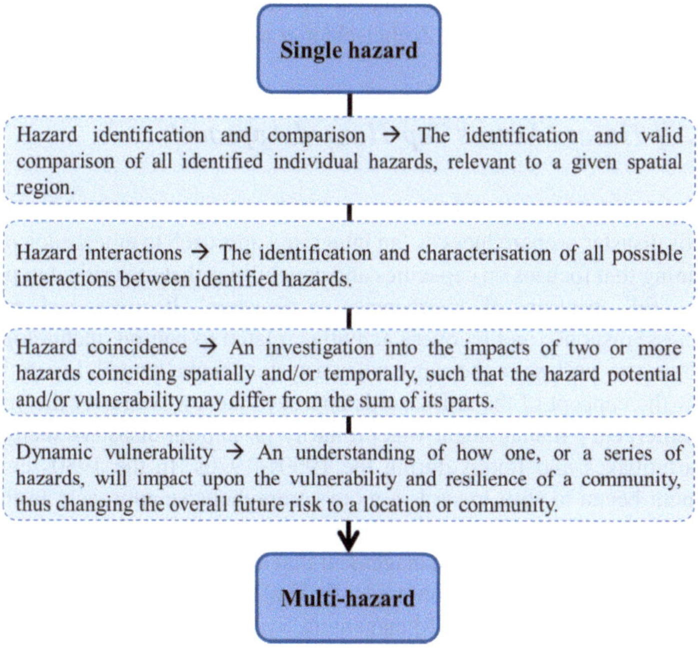

Fig. 3.1 Progression from a single-hazard approach to a multi-hazard approach (Modified from Gill and Malamud 2016)

The aforementioned challenges exist within the context of a highly connected and urbanizing world in which the dynamism of single and multi-hazards is changing at an unprecedented rate. Transboundary risk assessment and management for multi-hazard risks across countries are still lacking but very much needed in this increasingly interconnected world. The interconnectedness of hazards and risks needs to be explicitly taken into account, both from the individual and system perspective, so that individual DRM measures do not influence systemic risks on other spatial scales in the future.

Therefore, some key challenges due to the complexities arising from low awareness of the multi-hazard approach are as follows:

- Diverse language on multi-hazard risk and a lack of overview of existing methods and tools (Kappes et al. 2012; Gallina et al. 2016)
- Lack of a clear framework and guidelines for multi-hazard risk assessment and management (Scolobig et al. 2017; De Ruiter et al. 2020)
- Poor understanding of dynamic feedback between hazard, exposure and vulnerability (Gill and Malamud 2016)
- Focus on single hazards under current conditions without focusing on multi-hazard risk interactions or future scenarios (Gallina et al. 2016)
- Assessment of only a few studies on the effectiveness of DRM measures across hazards, sectors and time horizons (Zaghi et al. 2016)
- Distinct lack of in-depth case studies on multi-hazard risk assessment and management (Gill et al. 2020; Zscheischler et al. 2018)

3.2.2 *All-Hazard Versus Top-Hazard Approach*

The United States Center for Medicare and Medicaid Services (CMS 2019) defines all-hazards disaster preparedness as 'an integrated approach to emergency preparedness planning that focuses on capacities and capabilities that are critical to preparedness for a full spectrum of emergencies or disasters'. It covers risk reduction, preparedness, response and recovery activities related to natural, technological and societal issues regardless of the cause when a system is challenged. Figure 3.2 illustrates how the concept of the All-hazards approach has evolved over the years. The focus of emergency management was primarily on natural disasters such as hurricanes, earthquakes and floods during the 1960s–1970s. In the 1980s emergency management began to shift towards a more comprehensive approach. In the 1990s the National Response Framework (NRF) was developed, which established an all-hazards approach to emergency management that applies to all levels of government and non-governmental organizations under Federal Emergency Management Agency (FEMA). The National Incident Management System (NIMS) was established by FEMA, which provides a common framework for all responders and stakeholders to work together during emergencies. In the 2010s, the all-hazards approach continued to evolve with the development of the SFDRR and the inclusion of Natural Hazards

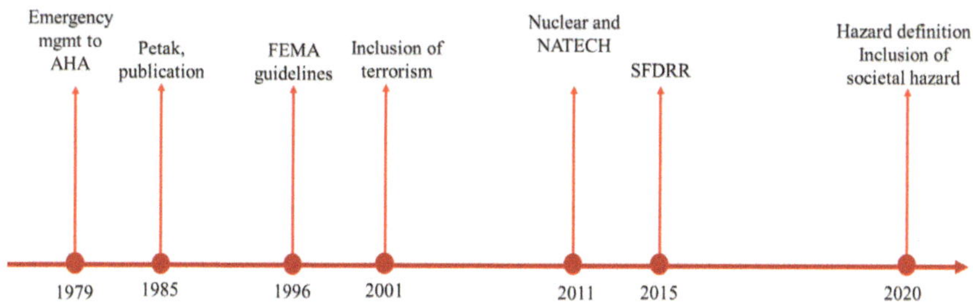

Fig. 3.2 Timeline of all-hazard approach

Triggering Technological Disasters (NATECH) and nuclear risks. It also emphasized the importance of engaging and involving all members of the community in emergency management efforts. In the 2020s after the COVID-19 pandemic, the importance of an all-hazards approach to emergency management that incorporates public health considerations was highlighted and societal hazards were included (CDC 2020).

Top-Hazards Approach (THA) prioritizes the hazard risks according to the local scenario. The approach suggests dealing with the top-hazard risks first to achieve better levels of preparedness. THA identifies with the fact that hazard risks and consequences are quantitatively and qualitatively different, and must not be consolidated in the process of disaster planning. New approaches in disaster planning should focus on the needs of the final beneficiaries more than the hazard type (Donahue et al. 2012) and disaster planning should start with the needs of a diverse population (gender, age, socio-economic, disability, etc.) that are associated with specific hazards.

Bodas et al. (2020) argue that although the 'All-Hazards Approach' (AHA) may present several arguable benefits, it has a risky self-destruct mechanism: 'grasp all, lose all'. A single investment in a generic set of solutions will achieve sufficient preparedness for all hazards and this potentially excuses decision-makers from giving the necessary attention to adequate disaster planning for higher-priority hazards. This is why they argue that instead of AHA disaster planners should employ THA. The focal priority of AHA is 'hazard', whereas for THA it is 'risk' (please see Table 3.2 for a comparative analysis).

However, similar to any framework in DRM, no approach is limitations-free. Firstly, a strong public reaction to what experts may perceive as minor risks can dramatically affect decision-making and skew the planning process (Kasperson et al. 1988). Disaster planners should optimize political interference in risk analysis processes while keeping the interests of all different stakeholders (Shrader-Frechette 1990). Secondly, hazards occurring less frequently will likely score lower during the prioritization process however their intensity will determine the degree of damage. Therefore, while THA is adapted to local context and circumstances, hazards with long return periods must be taken into consideration. Papadopoulos et al. (2014) cited an apt example in this regard where the risk of strong tsunamis in the Mediterranean Sea occurs roughly every 90 years. Finally, low-risk hazards can transition to

Table 3.2 Comparison of the principles of the top-hazards approach and the all-hazards approach (Bodas et al. 2020)

Item	All-hazard approach	Top-hazard approach
Main focal point	Hazards	Risks
Primary premise	Different hazard scenarios share certain commonalities and therefore should be managed with a common plan for hazard mitigation and preparedness	Since risks are quantitatively and qualitatively different, they must not be consolidated in the process of disaster planning
Approach in a nutshell	Utilize a generic system of response that fits most (if not all) hazards based on shared commonalities across hazards	Prioritize top-ranking hazards according to their risk and resolve them first with a comprehensive response plan
Main argument for	Cost effective for disaster planning	A robust framework for better preparedness
Main argument against	False sense of readiness	Leaves out certain risks unattended

high-risk ones in the absence of inadequate resources for resilience building. Thus, prioritizing top risks over low-risk ones can lead disaster planners to neglect such risks and might mislead them towards ranking such risks lower on the prioritization scale. This highlights the need to ensure that the risk analysis process is dynamic, in absence of which, THA may cause overlooked risks to become emerging high-risk threats.

3.3 Pathway for Holistic Management of Risks

The International Federation of Red Cross and Red Crescent Societies (IFRC) global guidelines note that having a common DRM plan can help ensure that resources are used efficiently and effectively, and can help build trust and cooperation among different stakeholders. This is corroborated by United Nations Office for Disaster Risk Reduction (UNDRR) which mentions the importance of having a common framework for DRM that can be adapted to the specific needs and context of different regions and hazards. A common adaptable plan is useful because it provides a systematic approach to managing disasters and emergencies, regardless of their cause or nature. This is followed in the all-hazards approach which consolidates several plans that an entity, like the government, might have in place to respond to different threats. It is advantageous as instead of the government demanding a specific plan for every scenario, the all-hazards approach provides a general framework on which all of the government's emergency management operations can be based. Additionally, it also supports the analysis of interlinked hazards.

Managing risks holistically for resilience building involves a comprehensive and integrated approach to DRM, which encompasses all phases of disaster management.

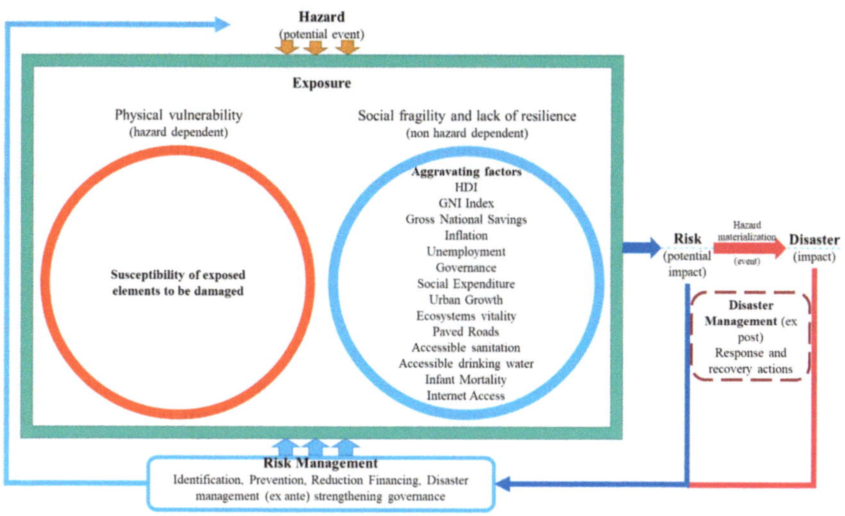

Fig. 3.3 Outline for holistic risk management (Modified from Marulanda-Fraume et al. 2021)

Figure 3.3 illustrates a holistic approach to risk management being adopted in recent times. The holistic approach states that to reduce existing risk or to prevent the generation of new risk the risk management system should be based on an institutional structure accompanied by the implementation of policies and strategies to intervene. This requires the participation of stakeholders in all stages of risk management.

Here are some key steps that decision-makers and other stakeholders can take to manage disasters holistically and build resilience:

- Establish a policy and legal framework: Decision-makers can start by establishing a policy and legal framework for DRM that defines roles and responsibilities, outlines funding mechanisms and provides a framework for coordination and cooperation among different stakeholders.
- Conduct risk assessments: Local governments can conduct risk assessments to identify hazards, vulnerabilities and capacities at the community level, and to inform preparedness and response planning.
- Develop preparedness and response plans: Local governments can work with communities to develop preparedness and response plans that are tailored to local needs and capacities, and that involve all stakeholders in the planning process.
- Strengthen early warning systems: Disaster management authorities can invest in early warning systems and technologies, such as weather monitoring and prediction systems, to provide timely and accurate information to communities and decision-makers.
- Build community resilience: Local governments, NGOs and CBOs can work with communities to build resilience through education and training, capacity building and the development of community-based disaster plans.

- Support recovery and reconstruction: Decision makers can support recovery and reconstruction efforts by addressing the underlying social, economic and environmental factors that contribute to risk, and by prioritizing the needs of the most vulnerable communities.
- Monitor and evaluate: Decision makers can monitor and evaluate DRM efforts to identify strengths and weaknesses and to ensure that DRM efforts are meeting the needs of communities and building resilience over time.

Overall, managing disasters holistically for resilience building requires a coordinated and integrated approach that involves all stakeholders. Decision makers, local governments, allied departments, NGOs, CBOs, etc. play a critical role in establishing policies and frameworks, investing in early warning systems and technologies, and working with communities to build resilience and address underlying risk factors. By taking these steps, stakeholders can help build more resilient communities and reduce the impact of disasters over time.

3.4 Discussion

This chapter attempts to give an insight of the length and breadth of the conceptualization and application of DRM and its linked factors important for its evolution. By considering this, various stakeholders in the field of DRM can gain the advantage of previous experiences and avoid repeating past mistakes in the future.

This chapter shows that the evolution of the risk management approach began with the initiation of civil defense in the mid-twentieth century, and this subsequently led to the formation of the concept of DRM during the 1990s, wherein the first steps for international collaboration were taken. The UN's three important conferences intended for disaster reduction played a special role in fostering cooperation among global communities. These scientific meetings, under the title, 'The World Conference on DRR', were hosted in three cities of Japan, namely, Yokohama (1994), Kobe (2005) and Sendai (2015), which subsequently led to the development of important global documents having set goals and objectives acting as a yardstick for decision-makers, and therefore, considered a breakthrough in the history of the risk management approach.

Following this international event, United Nations Office for Disaster Risk Reduction (UNISDR) was launched in 1999 to prepare countries to respond to disasters. Initially, the provision of an appropriate response to disasters was the main concern for countries, but this approach gradually changed, and pre-disaster events started to be prioritized rather than disaster response and post-disaster events, which were more focused on in the past. It is realized that in the risk management approach, disaster risk is to be managed rather than the disaster alone. Therefore, instead of thinking of the response to a disaster, the whole process of a risk management cycle is to be considered. With this approach, the residual risk can be reduced, and subsequently, the response mechanism will be smaller and less costly.

Referring to the COVID-19 pandemic, McClean (2020) stated that COVID-19 shed new light on the systemic nature of risk as seen today. A public health disaster quickly became a socio-economic disaster on a scale that had never been seen before. This worst tragedy of the twenty-first century taught us to strengthen disaster risk governance to deal with the threats which resulted in damage to the physical, economic and social well-being of millions of people. Hence, that is why governments need to develop their national DRR strategies with a clear understanding of the systemic nature of disaster risk.

3.5 Conclusion

An effective DRM approach involves a combination of strategies. In present times, risk management is recognized as an indispensable component of sustainable development, and efforts are ongoing to integrate DRM into all facets of development planning to shape resilient communities, countries, and regions. Various challenges in the present risk management approaches include limited awareness of a clear framework and guideline for multi-hazard risk assessment, low priority to beneficiary needs and so on. All-hazards approach has been a widely used framework but has some inherent weaknesses such as a lack of consideration for uneven hazard distribution that expose disaster managers to serious challenges in implementation. The top hazard approach overlooks the interconnectedness of hazard risks while prioritizing one. Thus, the 'best' DRM approach can be considered as the one that is tailored to the specific needs and circumstances of a given community or region and promotes actions related to risk assessment and planning, coordinated efforts for rescue and response, robust early warning systems, preparedness measures, incorporation of new technologies such as remote sensing and Geographic Information Systems (GIS) and integrating community knowledge into decision-making processes.

References

Adger WN (2000) Social and ecological resilience: are they related? Prog Hum Geogr 24(3):347–364

Bodas M, Kirsch TD, Peleg K (2020) Top hazards approach—rethinking the appropriateness of the all-hazards approach in disaster risk management. Int J Disaster Risk Reduct 47:101559

CATDAT (2021) Worldwide natural catastrophes loss database, Karlsruhe, Germany. http://www.risklayer.com/de/service/catdat/. Assessed 27 Jan 2022

CDC (2020) Interim guidance for Emergency Medical Services (EMS) systems and 911 Public Safety Answering Points (PSAPs) for COVID-19 in the United States. Centers for Disease Control and Prevention

CMS (2019) Centers for medicare and medicaid services. Emergency preparedness—updates to appendix Z of the state operations manual. https://www.cms.gov/Medicare/Provider-Enrollment-and-Certification/SurveyCertificationGenInfo/Downloads/QSO19-06-ALL.pdf. Assessed 27 Jan 2022

CRED (2021) Centre for research on the epidemiology of disasters. Disaster year in review 2020, global trends and perspectives. CRED & UC Louvain, Belgium. https://reliefweb.int/sites/reliefweb.int/files/resources/CredCrunchNewsletter,IssueNo.62(May2021)-DisasterYearinReview2020-GlobalTrendsandPerspectives.pdf. Assessed 27 Jan 2022

Cutter SL, Barnes L, Berry M, Burton C, Evans E, Tate E, Webb J (2003) A place-based model for understanding community resilience to natural disasters. Glob Environ Change 13(1):1–12

De Ruiter MC, Couasnon A, Van den Homberg MJC, Daniell JE, Gill JC, Ward PJ (2020) Why we can no longer ignore consecutive disasters. Earths Future 8:e2019EF001425. https://doi.org/10.1029/2019EF001425

Donahue DA, Cunnion SO, Balaban CD, Sochats K (2012) The all needs approach to emergency response. Homel Secur Affairs 8(1):1–17

Gallina V, Torresan S, Critto A, Sperotto A, Glade T, Marcomini A (2016) A review of multi-risk methodologies for natural hazards: consequences and challenges for a climate change impact assessment. J Environ Manag 168:123–132. https://doi.org/10.1016/j.jenvman.2015.11.011

Gill JC, Malamud BD (2016) Hazard interactions and interaction networks (cascades) within multi-hazard methodologies. Earth Syst Dynam 7:659–679. https://doi.org/10.5194/esd-7-659-2016

Gill JC, Malamud BD, Barillas EM, Guerra Noriega A (2020) Construction of regional multi-hazard interaction frameworks, with an application to Guatemala. Nat Hazards Earth Syst Sci 20:149–180. https://doi.org/10.5194/nhess-20-149-2020

Hewitt K, Burton I (1971) The hazardousness of a place: a regional geology of damaging events, vol 5. Department of Geography Research Publication, University of Toronto, Toronto, Canada, pp 154–155. ISBN 0802032818

IPCC (2018) Intergovernmental panel on climate change. Global warming of 1.5 °C: an IPCC special report on the impacts of global warming of 1.5 °C above pre-industrial levels and related global greenhouse gas emission pathways, in the context of strengthening the global response to the threat of climate change

IPCC (2022) Intergovernmental panel on climate change. Climate change 2022, impacts, adaptation and vulnerability, summary for policymakers. In: Working group II contribution to the sixth assessment report of the IPCC. Cambridge University Press, Cambridge. https://www.ipcc.ch/report/ar6/wg2/. Assessed 27 Jan 2022

Kappes MS, Keiler M, Von Elverfeldt K, Glade T (2012) Challenges of analyzing multi-hazard risk: a review. Nat Hazards 64:1925–1958. https://doi.org/10.1007/s11069-012-0294-2

Kasperson RE, Renn O, Slovic P, Brown HS, Emel J, Goble R, Kasperson J, Ratick S (1988) The social amplification of risk: a conceptual framework. Risk Anal 8(2):177–187

Kelman I, Gaillard JC, Lewis J, Mercer J (2016) Learning from the history of disaster vulnerability and resilience research and practice for climate change. Nat Hazards 82:129–143. https://doi.org/10.1007/s11069-016-2294-0

Marulanda-Fraume P, Cardona OD, Marulanda MC, Carreño ML (2021) Evaluating risk from a holistic perspective to improve resilience: a subnational level evaluation in Colombia. Revista Internacional de Desastres Naturales, Accidentes e Infraestructura Civil 19(20):1

McClean D (2020) Nature-based solutions vital for risk governance. United Nations Office for Disaster Risk Reduction (UNDRR). https://www.undrr.org/news/nature-based-solutions-vital-risk-governance. Assessed 27 Jan 2022

Papadopoulos GA, Gràcia E, Urgeles R, Sallares V, De Martini PM, Pantosti D, Papageorgiou A (2014) Historical and pre-historical tsunamis in the Mediterranean and its connected seas: geological signatures, generation mechanisms and coastal impacts. Mar Geol 354:81–109

Rajabi E, Bazyar J, Delshad V, Khankeh HR (2022) The evolution of disaster risk management: Historical approach. Disaster Med Public Health Prep 16(4):1623–1627

Scolobig A, Komendantova N, Mignan A (2017) Mainstreaming multi-risk approaches in policy. Geosciences 7:129. https://doi.org/10.3390/geosciences7040129

Shrader-Frechette K (1990) Scientific method, anti-foundationalism, and public decision making. Risk 1:23

UNCED (1992) Agenda 21. United Nations, New York. https://sustainabledevelopment.un.org/content/documents/Agenda21.pdf. Assessed 27 Jan 2022

UNDRR and WMO (2022) Global status of multi-hazard early warning systems: Target G. United Nations Office for Disaster Risk Reduction

UNDRR (2007) Hyogo framework for action 2005–2015: building the resilience of nations and communities to disasters. https://www.undrr.org/publication/hyogo-framework-action-2005-2015-building-resilience-nations-and-communities-disasters. Assessed 15 Nov 2023

UNDRR (2015) Sendai framework for disaster risk reduction 2015–2030. UNDRR, Geneva. https://www.preventionweb.net/files/43291_sendaiframeworkfordrren.pdf. Assessed 27 Jan 2022

Ward PJ, Daniell J, Duncan M, Dunne A, Hananel C, Hochrainer-Stigler S, De Ruiter MC (2022) Invited perspectives: a research agenda towards disaster risk management pathways in multi-(hazard-) risk assessment. Nat Hazard 22(4):1487–1497

WHO (2020) World Health Organization. WHO director-general's opening remarks at the media briefing on COVID-19. World Health Organization

Wisner B, Blaikie P, Cannon T, Davis I (2004) At risk: natural hazards, people's vulnerability, and disasters. Routledge

Zaghi AE, Padgett JE, Bruneau M, Barbato M, Li Y, Mitrani-Reiser J, McBride A (2016) Establishing common nomenclature, characterizing the problem, and identifying future opportunities in multi-hazard design. J Struct Eng 142:H2516001. https://doi.org/10.1061/(ASCE)ST.1943-541X.0001586

Zscheischler J, Westra S, Van den Hurk BJJM, Seneviratne SI, Ward PJ, Pitman A, AghaKouchak A, Bresch DN, Leonard M, Wahl T, Zhang X (2018) Future climate risk from compound events. Nat Clim Change 8:469–477. https://doi.org/10.1038/s41558-018-0156-3

Zschau J (2017) Where are we with multihazards, multirisks assessment capacities? Global status of multi-hazard early warning systems Target G; pp 98–130. Access on 03.06.2024 at https://drmkc.jrc.ec.europa.eu/portals/0/Knowledge/ScienceforDRM/ch02/ch02_subch0205.pdf

Part II
Different Types of Risks

Chapter 4
Adapting All-hazards Planning to Chemical, Biological, Radiological, and Nuclear (CBRN) Threats

R. Alexander Hamilton and John Jones

Abstract Attention to the harm posed by chemical, biological, radiological, and nuclear (CBRN) events is growing. Informed by events ranging from COVID-19 to the 2020 Beirut explosion, it is apparent that events of this kind, whether natural, accidental, or deliberate in origin, have the potential to cause serious disruption to the functioning of a society, causing widespread human, material, economic, or environmental losses. In other words, CBRN events can reach the level of (human-induced) disasters. Despite this observation, such events, and especially deliberate use (terrorism) scenarios (CBRN threats), tend to be treated as a distinct category of disaster and often do not, at least explicitly, figure in the disaster risk reduction literature, which has historically focused on natural hazards. In this chapter, we consider to what extent CBRN threats justify distinct categorization and exceptional planning. We argue that, while CBRN threats introduce a number of distinct challenges (due to the need to account for the active role played by a perpetrator), many aspects of the risk management cycle apply to all disasters, irrespective of their origin. We further argue that generic 'all-hazards' plans provide a suitable basis for addressing CBRN threats, albeit such plans must necessarily be complemented by provisions adapted to this distinct kind of human-induced disaster.

Keywords Chemical · Biological · Radiological · Nuclear · CBRN · Threat · Terrorism · Disaster · Risk · Resilience · All-hazards · Prevention · Detection · Response

---

The original version of the chapter has been revised: Chapter title has been updated. A correction to this chapter can be found at
https://doi.org/10.1007/978-981-97-1860-3_19

R. Alexander Hamilton (✉)
United Nations Interregional Crime and Justice Research Institute (UNICRI), Turin, Italy
e-mail: alexander.hamilton@un.org

J. Jones
Hampshire, UK

T. Izumi et al. (eds.), *All-Hazards Approach*, Disaster Risk Reduction,
https://doi.org/10.1007/978-981-97-1860-3_4

4.1 Introduction

Attention to the harm posed by chemical, biological, radiological, and nuclear (CBRN) events is growing. From the COVID-19 pandemic, caused by the global spread of a novel coronavirus (a biological agent), to the 2020 Beirut explosion, triggered by the ignition of ammonium nitrate (a chemical compound used in fertilizers and explosives) improperly stored at the Port of Beirut, it is apparent that events of this kind can cause serious disruption to the functioning of a society, causing widespread human, material, economic, or environmental losses. In other words, such events can reach the level of (human-induced) disasters.[1] Yet, references to 'CBRN', and especially potential deliberate use (terrorism) scenarios (CBRN threats), do not figure prominently in the disaster risk reduction literature, which has historically focused on natural hazards (Frulli 2022). Indeed, even in the case of the *Sendai Framework for Disaster Risk Reduction 2015–2030* (Sendai Framework), which has been commended for broadening the scope of the disaster risk reduction definition "to focus on *both natural and man-made* hazards and related environmental, technological and biological hazards and risks" (Sendai Framework, Forward, emphasis added), CBRN threats are not explicitly mentioned.

Because disaster risk reduction finds its origins in national, regional, and global efforts to address familiar natural hazards, such as typhoons and earthquakes, it is not surprising that references to CBRN threats (a far less common type of event, albeit potentially no less destructive) are limited in the disaster risk reduction literature. Even within the framework of global processes dedicated to international security, attention to the deliberate use of CBRN materials by non-state actors (criminals and terrorist groups, as opposed to States) is relatively new, gaining traction especially since the signing of United Nations Security Council Resolution (UNSCR) 1540 (2004),[2] which obliges States to adopt measures to counter the proliferation and use of so-called weapons of mass destruction (WMD) by non-state actors. However, irrespective of the relative novelty of CBRN threats, it is important that governments, which increasingly embrace the 'all-hazards' approach,[3] are sensitive to a range of scenarios of this kind, and take adequate measures to prevent, detect, and prepare for such contingencies.

In this chapter, we take a closer look at CBRN disaster risks, with a special focus on CBRN threats, posing several key questions, including: To what extent are CBRN threats distinct from traditional (natural) disasters? What unique risk management considerations do CBRN threats introduce? And how can attention to

---

[1] Please see 'disaster' definition employed in the *ASEAN Agreement on Disaster Management and Emergency Response*, available at: https://agreement.asean.org/media/download/20220330063139.pdf.

[2] For more information on UNSCR 1540 (2004), please see: https://www.un.org/disarmament/wmd/sc1540/.

[3] For the purposes of this chapter, we have adopted Waugh's (2004, p. 12) definition of 'all-hazards': "… adaptable plans that provide the basis for dealing with a variety of hazards and disasters, including terrorism."

CBRN threats be effectively integrated into existing all-hazards risk management frameworks? We argue that, while CBRN threats (due to the need to account for the active role played by a perpetrator) pose a number of distinct challenges, many aspects of the risk management cycle (prevention, detection, response) apply to all disasters, irrespective of their origin. We further argue that generic 'all-hazards' plans provide suitable basis for addressing CBRN threats but must be complemented by provisions adapted to the distinct requirements presented by this specific kind of human-induced disaster.

4.2 Types of CBRN Events

In general, CBRN events occur because of the loss of control over, or the inability to contain, hazardous materials which in turn causes the spread of contamination (in the broadest sense of the word) to people and to the environment. Distinguishing between different types of CBRN events is primarily a matter of accounting for *how* the hazardous material (a toxic chemical compound, a pathogen, or a radioactive substance) is released (see Table 4.1). Some CBRN events can be triggered by natural occurrences such as extremes of weather that result in containment failures or through the evolution of a virus. Others may be accidental—in other words there is no intention to cause harm but there has been a system failure due to improperly functioning equipment and/or human error. Finally, CBRN events can be deliberate, the result of an intentional act by an individual or group of individuals. For those working in security fields (e.g., law enforcement), the acronym 'CBRN' (or 'CBRNE', adding explosives) is generally reserved for this type of event. The acronym signifies an envelope of scenarios involving the potential deliberate release of a hazardous CBRN material by a perpetrator (a criminal or terrorist actor) who possesses the motive and capabilities (concepts we will return to in this section) to execute a specific attack. While those working in security fields are no less cognisant that CBRN materials can be released by natural (e.g., emerging infectious diseases) or accidental (e.g., industrial accidents) means, the CBRN acronym in security contexts is generally understood to signify a deliberate (terrorist) act. In this section, we briefly review these three types of CBRN events, drawing attention to what makes CBRN threats distinct.

4.2.1 *Natural CBRN Events*

Naturally occurring CBRN events are most commonly associated with the natural occurrence of infectious diseases, causing illness in humans, animals, or plants. Pandemics, such as COVID-19, can originate at any time, most often through the transmission of a naturally circulating virus that has undergone one or more mutations that have enabled it to jump the species barrier between animals and humans. This

Table 4.1 Different types of CBRN events

Types of CBRN events			
	Natural	Accidental	Deliberate
Description	Natural occurrence of infectious disease, causing illness in humans, animals or plants, or natural disasters triggering uncontrolled release of hazardous material from a CBRN facility, transport or storage site	Unintended release of hazardous material from a CBRN facility, transport or storage site due to mechanical failure, human error, negligence or a combination of two more of these elements	Intentional release of CBRN material by a State (warfare) or non-state (terrorism) actor possessing specific motives (e.g. political or ideological beliefs) and capabilities(e.g. financial resources and technical know-how)
Examples	COVID-19 pandemic (2020–2023)	Beirut explosion (2020)	Anthrax letter attacks (2001)
	West Africa Ebola virus epidemic (2013–2016)	Graniteville train crash (2005)	Tokyo Subway Sarin attack (1995)
	Fukushima Daiichi Nuclear Accident (2011)	Chernobyl nuclear disaster (1986)	Rajneeshee bioterror attack (1984)
	SARS outbreak (2002–2004)	Bhopal chemical disaster (1984)	WWI chemical weapons use (1914–1918)

process, known as a zoonotic disease event, has been the primary cause of pandemics throughout human history, as well as the primary cause of major epidemics, ranging from the 2002–2004 SARS outbreak to the 2013–2016 West Africa Ebola virus epidemic. Due to the severity of pandemics, both in terms of the loss of human life and the economic ramifications of global lockdowns, this type of event typically ranks among the top hazards countries face as assessed in national threat assessments (OECD 2018).

Other CBRN events in this category trace their origin to extreme weather events, such as typhoons and earthquakes, which can cause the uncontrolled release of hazardous material from a CBRN facility, transport, or storage site, resulting in environmental contamination and possible exposure to human populations. Also referred to as Natech (Natural Hazards Triggering Technological Disasters) accidents,[4] these events exist at the intersection of natural disasters and industrial accidents, introducing significant complexity and potentially having far-reaching consequences (OECD 2022). The 2011 Fukushima Daiichi Nuclear Accident represents an extreme case of a Natech accident, resulting from an undersea earthquake that triggered a tsunami that in turn damaged the Fukushima Daiichi nuclear power station, causing the release of radioactive contamination into the environment, ultimately

[4] For an online repository of Natech accident cases, please see: https://enatech.jrc.ec.europa.eu/list/natech.

resulting in the worst accident at a nuclear power plant since the Chernobyl disaster in 1986.[5]

4.2.2 Accidental CBRN Events

Accidental CBRN events involve the unintended release of hazardous material from a CBRN facility, transport, or storage site, whether due to mechanical failure, human error, negligence, or a combination of two or more of these elements. Events of this kind range from highly complex emergencies such as Chernobyl and Bhopal (the world's worst industrial chemical disaster) to far more frequent, smaller-scale events such as transport accidents involving toxic chemicals. Significantly, while these events are by definition unintended or accidental, rather than deliberate, the role individuals play in precepting or aggravating a given crisis, either due to human error or (criminal) negligence, is often a key factor. For example, while the immediate cause of the 2020 Beirut explosion was the unintended ignition of large amounts of ammonium nitrate stored at a warehouse at the port of Beirut, allegations of "systemic problems of negligent governance and widespread corruption"[6] have been cited as key contributing factors. Flaws or lack of consistency in risk assessment can also play a role. The fact that 'nothing has gone wrong so far' is not a good indicator that nothing can go wrong in the future.

4.2.3 Deliberate CBRN Events

Deliberate CBRN events, by contrast, involve the *intentional* release of CBRN material by direct (e.g., aerosol delivery) or indirect (e.g., sabotage) means. Historically, a number of States have developed and deployed CBRN weapons during times of war (e.g., the use of chlorine gas during World War I). This legacy of use, in combination with the rise of large state-funded programs during the Cold War period, culminated in the signing of international conventions and treaties banning the development, storage, stockpiling, and use of CBRN weapons.[7] While the threat of CBRN weapons use by States remains a critical international security issue, concerns about non-state actors (criminals and terrorist groups) have grown in recent years. In particular, following the collapse of the former Soviet Union and a number of high-profile terrorist attacks, including the Tokyo Subway sarin attack in 1995, and especially

---

[5] For further information on the Fukushima Daiichi Nuclear Accident, please see: https://www-pub.iaea.org/MTCD/Publications/PDF/Pub1710-ReportByTheDG-Web.pdf.

[6] Please see: https://www.ohchr.org/en/press-releases/2022/08/un-experts-call-international-investigation-2020-beirut-explosion.

[7] Notably, the Nuclear Non-Proliferation Treaty (NPT), signed in 1968, the Biological Weapons Convention (BWC), signed in 1972, and the Chemical Weapons Convention (CWC), signed in 1993.

since the 2001 anthrax letter attacks, growing attention has been paid to the 'non-state threat' (Collier et al. 2004; Wright 2004; McLeish and Nightingale 2007; Hamilton 2018).

Key to understanding what makes CBRN threats (and here and throughout this chapter we use this term to refer to the deliberate use of CBRN materials by non-state actors, rather than States) distinctive is recalling that they are neither natural phenomena nor accidents. Rather, CBRN threats are mediated by an individual or group of individuals who possess specific motives (e.g., political or ideological beliefs) and capabilities (e.g., financial resources and technical know-how) to carry out a specific attack. It is this combination of 'motive' and 'capability' that analysts use to assess the relative significance and feasibility of CBRN threats. In general terms, motive can be understood as driving the pursuit of CBRN weapons, and capability as the means of developing and deploying them (Binder and Ackerman 2023). Fortunately, even among highly motivated and well-funded terrorist groups, the necessary capabilities to transform CBRN materials into effective weapons have proven highly limited, contributing to the limited frequency and impact of past events. Indeed, the vast majority of cases involving the deliberate use of CBRN materials by non-state actors have been smaller-scale events that while potentially "disruptive and psychologically potent" have not yielded "the casualty levels or physical destruction generally associated with a WMD" (Ackerman and Jacome 2018, p. 23).[8] Nevertheless, the growing availability of CBRN materials used for legitimate purposes, in combination with emerging technologies that have the potential to lower technical barriers to weapons development, represent important trends (among others) that have the potential to significantly alter the risk outlook in the years to come (ibid.; Binder and Ackerman 2023).

As rational actors (Mair and Mair 2003), potential perpetrators will equally weigh the costs and benefits of their actions (or inactions), choosing between more or less easily obtainable options to fulfill their goals. In practice, this means that measures taken to make CBRN materials or weapons less obtainable or attractive to aspiring perpetrators can help limit the likelihood of potential attacks. Attention to these considerations provides an important starting point for crafting, in particular, prevention measures adapted to CBRN threats (a subject we will return to later in this chapter). While governments cannot foresee all possible CBRN threats, they can take steps to anticipate a range of more or less likely scenarios (based on understanding potential perpetrators' motives and capabilities) and introduce security measures that make CBRN acquisition more difficult (e.g., personnel reliability screening) or less attractive (e.g., surveillance and intelligence operations that make it more likely perpetrators will be discovered).

---

[8] Based on the University of Maryland's "Profiles of Incidents Involving CBRN by Non-state Actors" database, there have been 517 events, "including attacks, failed agent-use attempts, plots and proto-plots", involving CBRN materials between 1990 and 2016 (https://www.start.umd.edu/research-projects/profiles-incidents-involving-cbrn-and-non-state-actors-poicn-database). With the possible exception of Aum Shinrikyo's dispersal of sarin on the Tokyo subway in March 1995 (killing 12 and injuring more than 1,000), among several other cases of uncertain origin, these events have largely fallen short of WMD-scale attacks (Ackerman and Jacome 2018).

4.3 Characterizing CBRN Threats

As we have seen, CBRN events, whether natural accidental or deliberate in origin, result in the release of hazardous materials that pose a risk to health and the environment. Distinctions between different types of CBRN events can be made based on how the material in question is released. In the case of CBRN threats, the fundamental attribute that makes this type of event distinct is the active role played by a perpetrator who chooses, based on the opportunities available to them and the capabilities they possess, how the event will manifest with a view to achieving specific goals. In the case of a natural disaster, or even an accident caused through inattention to procedures or negligence, by contrast, there is no design that drives the choice of hazardous material; no rational actor that chooses between release options or seeks to anticipate, misdirect, or undermine response efforts. The introduction of a rational and determined adversary, the hallmark of CBRN threats, does introduce these features, adding intentionality and a further degree of complexity into the risk calculus.

While it is beyond the scope of this chapter to consider at any length the host of motivations (mass casualties, economic damage, financial gain, etc.) that have been attributed to non-state actors who have pursued CBRN weapons or to elaborate on the capabilities and constraints (financial, technical, operational, etc.) of would-be proliferators, there is value in considering a range of potential CBRN threat scenarios. In doing so, we are able to develop a clearer picture of what CBRN threats can look like, shedding light on potential vulnerabilities and gaps in disaster risk reduction plans as well as opportunities for constructive and proactive risk mitigation actions. With this in mind, in the following we present one set of CBRN threat scenarios drawn from the International Atomic Energy Agency (IAEA) publication, *Developing a National Framework for Managing the Response to Nuclear Security Events* (IAEA 2019, p. 11), a publication that provides guidance to States on the development, implementation, maintenance, and sustainment of a national framework for managing the response to nuclear security events:

- "Sabotage of a nuclear facility or nuclear material resulting in a release of energy and/or dispersal of radioactive material.
- Sabotage of an associated facility using or storing radioactive material or an associated activity (e.g. transport of radioactive material) resulting in dispersal of radioactive material.
- Operation of a radiation exposure device (RED) or series of REDs, such as a high activity radioactive source, exposing people in its vicinity to radiation.
- Operation of a radiological dispersal device (RDD) or series of RDDs resulting in dispersal of radioactive material by means of explosives or other means of dispersal (e.g. an aerosol generator, via a building ventilation system, manually).
- Dispersal of nuclear material or energy release (and dispersal of radioactive material) from a detonation caused by a fission chain reaction in nuclear material.
- Introduction of radioactive contamination at or into one of the following:

- A strategic location, such as the venue of a major public event;
- The food chain;
- The water supply network;
- Cosmetic, pharmaceutical or other products used by the public."

While the above examples are specific to the case of radioactive and nuclear materials, they illustrate a range of contingencies whereby non-state actors could exploit vulnerabilities in systems designed to prevent unauthorized access to hazardous CBRN materials in general. They also showcase how such materials could be dispersed with the intent to cause harm. It must nonetheless be emphasized that, with scenarios ranging from the sabotage of a nuclear facility to the dispersal of radioactive material using conventional explosives (a 'dirty bomb') to the detonation of a nuclear weapon, these examples equally illustrate a vast difference in technical sophistication, capturing events that range significantly in their likelihood and impact. In general, all things being equal, the more sophisticated the weapon or attack modality the less likely and more damaging its use or application. With this, and recalling that historical cases of non-state CBRN events demonstrate a tendency toward less sophisticated attacks, planners must necessarily adjust their assessments to match the capabilities of potential adversaries. For example, careful assessments would note that "the release of in situ toxic industrial chemicals or highly radiological materials close to an urban area" is a far more plausible and potentially concerning near-term threat than "the highly unlikely serendipitous acquisition of a viable nuclear or biological weapon from a state arsenal" (Ackerman and Jacome 2018, p. 34).

Having considered a range of CBRN threat scenarios, it is worthwhile to take a step back to revisit some of the distinct characteristics of deliberate use events in general, touching on the implications for planners and responders who may be more familiar with natural and accidental hazards, which do not involve the active participation of a perpetrator. Some key characteristics of CBRN threats and their implications include the following:

- The selection of materials that are highly hazardous, including combinations of materials to augment the overall impact;
- The selection of targets (facilities, storage sites, or transports that could be a source of material or sites of sabotage) based on perceived vulnerabilities;
- The selection of delivery systems or dissemination methods to maximize the possibility of exposing specific groups or communities;
- Clandestine or covert attacks, including the release of materials that are unfamiliar, not easily identified, not easily attributable, and/or result in the delayed onset of symptoms, delaying the initiation of response operations;
- The use of disinformation to actively mislead response operations and/or to generate heightened uncertainty or fear to augment perceived risk.

4.4 CBRN Threats and the Risk Management Cycle

Like other unconventional hazards, CBRN events, due their relative complexity, infrequency, and possible far-reaching consequences, require flexible and adaptive approaches to risk mitigation and management. While true of CBRN events in general, this is especially true in the case of CBRN threats due to the role of potential perpetrators who will seek to actively design and craft conditions in favor of achieving specific goals. In this section, we explore how prevention, detection, and response efforts can be adapted to CBRN threats.

4.4.1 Prevention

Unlike natural hazards, which offer few opportunities to prevent events from happening (we can model weather patterns, but we cannot prevent storms), CBRN threats (due to the active and necessary role played by perpetrators) present multiple intervention points for preventative action. In general, these interventions seek to make it more difficult for potential perpetrators to exploit vulnerabilities, acquire and/or misuse CBRN materials. While by no means an exhaustive list, some of the primary interventions of this kind are as follows:

- Physical protection measures (CCTV surveillance, security guards, fences, locking doors, etc.): seek to reduce unauthorized access to toxic, pathogenic, or radioactive substances, including during the production, use, storage, transportation, sale, and disposal of these materials;
- Personnel reliability screening (security, medical, and psychological evaluation programs): seek to ensure only trusted and reliable individuals are granted access to hazardous materials and sensitive sites and information, recognizing that 'insiders' can directly or indirectly misuse privileged access and knowledge;
- Intelligence and law enforcement activities (human-led intelligence and law enforcement operations): seek to monitor known groups or individuals of concern and gather intelligence about emerging threats in an effort to anticipate and pre-empt CBRN weapons proliferation and attacks;
- Inventory management systems (e.g., electronic databases detailing the types, quantities, and storage locations of hazardous materials): seek to accurately document and account for hazardous materials during their use, storage, transportation, or sale to ensure they are not lost, stolen, or diverted for malicious use;
- Security cultures (codes of conduct, awareness raising, education, and training programs, etc.): seek to sensitize scientists, researchers, and technicians about potential threats, vulnerabilities, and opportunities for preventative action;
- Engagement with the private sector and the general public (outreach and awareness raising programs): seek to establish a baseline level of awareness about potential threats and encourage non-specialists to report suspicious behaviors or activities.

While further prevention measures exist, including actions that can be taken to help prevent terrorism in general such as programs dedicated to preventing radicalization, the above examples illustrate a variety of interventions that can help countries establish or strengthen prevention programs adapted to CBRN threats. It is important to note that no single intervention is sufficient and implementing prevention measures involves a large number of government and non-government entities working in a coordinated way with industry and the public. Finally, it is also worth noting that, historically, significant emphasis has been placed on restricting access to hazardous materials, sensitive information and so-called 'dual-use' technologies (technologies that can be used for both productive and destructive purposes), motivating international and national regimes and frameworks dedicated to keeping such items out of the 'wrong hands'. In recent years, however, in light of the growing recognition that many items of this kind are already widely available (due to increased global trade and the evolution of information technology, among other trends), greater emphasis has been placed on shaping the behavior of those who work most closely with the very materials, information, and technologies of greatest concern. This has motivated complementary initiatives, programs, and activities at the international and national levels aimed at sensitizing practitioners (above all scientists, researchers, and technicians who work in CBRN fields) about potential threats and enlisting them to play an active role in anticipating and preventing misuse.

4.4.2 Detection

Detecting CBRN threats when or soon after they occur is critical to initiating a rapid and coordinated response that will save lives, mitigate environmental and economic damage, and ultimately restore normal living and working conditions. Because many CBRN materials are invisible and/or have latent effects, and perpetrators may seek to actively conceal an attack and/or to misdirect authorities, determining that an attack has even happened can be a non-trivial matter. Consequently, detection ideally begins well 'upstream' of a deliberate CBRN release, drawing on law enforcement and intelligence communities, working hand in hand with customs and border control authorities (among others), to identify and detect suspicious people, activities, materials, and technologies that could signal a potential future attack. Elements supporting this goal include, but are not limited to the following:

- Developing a detection strategy and systematic approach based on threat assessment to identify suspicious activity;
- Conducting education and training activities to sensitize and test relevant workforces to successfully lead and conduct intelligence-led operations;
- Investing in detection equipment and technologies to detect CBRN materials that fall outside of regulatory control, especially at points of entry, transportation hubs, major cities, and other potential targets such as major public events.

Significantly, relative to chemical, radiological, and nuclear events, which are likely to be accompanied by or cause relatively immediate and visible consequences (sabotage at a nuclear power plant triggering alarms, an explosive discharge associated with a radiological dispersal device, the rapid onset of symptoms following exposure to chemical agents, etc.), biological events tend to be far more difficult to detect in real-time. This is because most biological agents have a relatively long latency period (the time between exposure and the manifestation of symptoms) and many cause diseases that present nonspecific symptoms (fatigue, headache, malaise, etc.), especially at an early stage of the disease's progression. Consequently, detecting the covert release of a biological agent relies heavily on effective health surveillance and effective communication and reporting between relevant agencies, including public health, animal health, and law enforcement entities. In the event of an overt or publicized release of a biological agent, specialist sampling and identification techniques are required.

4.4.3 *Response*

CBRN events in general, and CBRN threats in particular, are complex and this complexity introduces particular challenges. Scene management, life-saving, and hazard-reduction activities have to take place simultaneously and under intense time-pressure. These challenges require coordinated situational awareness and effective inter-agency communication, demanding not only plans and planning, but also regular exercising to help ensure responding agencies understand how to perform if and when specific CBRN events happen.

In the United Kingdom (UK), the emergency services have developed a joint system for responding to CBRN terrorism events, detailed in the report, *Responding to a CBRN(e) Event: Joint Operating Principles for the Emergency Services* (2016, First Edition). The report presents a range of CBRN threat scenarios, underlining that "[t]he impact and response will vary depending upon the nature of the material and event, for example:

- A chemical attack may produce rapid onset of severe symptoms. Many chemical agents can be readily detected and potentially identified with specialist equipment.
- A biological release may not be identified for some time and may only be recognised through health monitoring. The scene of any release may be unidentified.
- A radiological release may be accompanied by explosives (a 'dirty bomb'), or the dispersal of radioactive particulates into the air, with no obvious sudden onset of symptoms.
- A nuclear attack is likely to be readily identified and result in immediate, catastrophic consequences and a long lasting radiation hazard.
- Explosives may be used as a means of dissemination for the above materials or, in its own right, as an additional method of attack." (ibid., p. 6)

The report continues:

> Whilst any such incident is ongoing there will be a continued threat to life. Uncertainty about release location(s), spread of contamination, intent and capability of the terrorists means that the management of life-saving interventions may be much more difficult than for other major incidents. The assumption will however be that once a CBRN(e) payload has been delivered, the perpetrators are unlikely to remain in the immediate vicinity, although they may present an ongoing threat in the wider area (ibid., p. 7).

In general, some of the particular challenges and requirements for responding agencies in the case of the deliberate release or attempted release of CBRN materials, include the following:

- A multi-agency command and control system needs to be in place to ensure a coordinated response;
- The ongoing presence of contamination at the scene requires additional protection for responders to prevent the inhalation or absorption of hazardous substances so that they do not become victims themselves;
- In the case of chemicals and some radioactive material, a timely response is vital so that any persons at the scene can be evacuated and receive medical attention, ideally within an hour of the initial release;
- The threat posed by the presence of perpetrators at access points to the scene who might wish to impede the responding agencies using firearms, explosive or incendiary devices may need to be neutralized before life-saving activities can begin.

Because the first responders to arrive at the scene are unlikely to be CBRN specialists, the initial operational response must also take some pragmatic steps to minimize loss of life until specialist, suitably protected, responders can arrive to deliver a full response. Several simple steps that can be taken to achieve this include the following:

- Improvised self-decontamination by those affected using available absorbent material;
- Designation of temporary safe areas upwind and away from the source of any contamination;
- Provision of information and reassurance to casualties;
- Joint on-scene command to oversee the initial management of the incident and the transition from the non-specialist to the specialist response.

These provisions, among others, can supplement generic all-hazards plans and can be complemented by the development of concepts of operation for specialist responders.

4.5 Adapting All-Hazards Planning to CBRN Threats: Conclusions and Recommendations

It is perhaps axiomatic that the event that occurs will not exactly match the event that was planned for. It is sometimes said, 'no plan survives first contact with the enemy'. To be prepared for CBRN threats, like any other disaster situation, planning is essential—a variety of scenarios must be considered, joint-procedures must be developed, exercises must be conducted—and capabilities—human, technical, operational—must be in place and ready to be activated and deployed. It is this understanding—that countries, businesses, communities, and individuals should have at their disposal a range of flexible and adaptive tools to apply to disasters if and when they occur—that underpins 'all-hazards' planning. However, as Waugh (2004, p. 12) rightly states, "'all-hazards' does not mean a perfect plan for every conceivable type of disaster. All-hazards means adaptable plans that provide the basis for dealing with a variety of hazards and disasters, including terrorism." Put another way, all-hazards plans do not need to, and cannot, account for all possible contingencies. Instead, they can and should provide a common basis for disaster response, offering a "unifying approach to dealing with hazards and disasters and a common terminology for emergency managers and public officials" (ibid., p. 11).

Due to the flexibility and cost-effectiveness of all-hazards emergency planning, and perhaps especially due to the growing need for governments "to cope with the complexity, novelty, ambiguity and uncertainty that characterise many modern crises" (OECD 2018, p. 15), the all-hazards approach has understandably emerged as a cornerstone of disaster risk reduction policy and practice. Importantly, for countries that have already adopted all-hazards plans, plans that may have been devised with a view to familiar natural hazards, the underlying principles communicated in these plans are no less applicable to CBRN events, including CBRN threats. An all-hazards "plan is the starting point and having generic evacuation, shelter-in-place, debris management, and other programs ready to be adapted to circumstances is far more efficient and understandable than stand-alone programs for each type of disaster" (ibid., p. 12). An all-hazards plan provides a basis for dealing with a range of disaster risks. This plan must then be complemented by risk- and threat-specific provisions.

In this chapter, we have presented a range of considerations and provisions specific to the case of preventing, detecting, and responding to CBRN threats. It is our hope that these can provide governments (in coordination with relevant stakeholders, from industry to academia) a starting point for adapting generic all-hazards plans to the potential deliberate use of CBRN materials. Ultimately, while CBRN threats present a number of distinct challenges (due to the active role of rational and determined adversaries who will seek to deliberately orchestrate crises), they are fundamentally another contingency that governments must plan and prepare for, drawing upon and, as needed, expanding existing resources and procedures. To this end, the all-hazards approach provides a suitable basis for addressing CBRN threats, among other (human-induced) disasters. It is not a question of whether the all-hazards approach is adequate or equipped to address deliberate use scenarios, but rather a question of how governments can work to successfully adapt generic plans to meet distinct requirements.

Finally, as it would seem, from our perspective, that the fields of disaster risk reduction and CBRN security have, to date, tended to operate in relative isolation, we believe that it is important to create further opportunities for these disciplines and their practitioners to work together more closely. While there are undoubtedly challenges to bridging the conceptual space between natural disasters and CBRN terrorism, these challenges are by no means insurmountable. The current publication is indeed an attempt at integrating diverse perspectives on disaster risk with a view to resilience building and cultivating a true 'all-hazards' approach. By sharing knowledge and experience, embracing an interdisciplinary approach, work of this kind helps to demystify our individual domains of expertise and helps build a common language for thinking about disasters (in the broadest sense of the word) and what we can do to address them. Moreover, this work helps clarify the many commonalities that exist between different types of disaster risks and the broad applicability of many risk management measures. There is therefore tremendous value in continuing to work together to share our collective experience to optimize rather than revolutionize the all-hazards approach.

References

Ackerman G, Jacome M (2018) WMD terrorism: the once and future threat. Prism 7(3):22–36

Binder MK, Ackerman GA (2023) CBRN terrorism. Published online 22 February 2023. https://oxfordre.com/internationalstudies/display/10.1093/acrefore/9780190846626.001.0001/acrefore-9780190846626-e-706;jsessionid=6FF5A087DBD5BF1B47F48D5FB2FAE358

Collier SJ, Lakoff A, Rabinow P (2004) Biosecurity: towards and anthropology of the contemporary. Anthropol Today 20(5):3–7

Frulli M (2022) The challenge of outlining the CBRN definitional framework. In: de Guttry A, Frulli M, Casolari F, Poli L (eds) International law and chemical, biological, radio-nuclear (CBRN) events. Brill | Nijhoff, Leiden. https://doi.org/10.1163/9789004507999_002

Hamilton RA (2018) Weaponization of infectious disease. In: Reichel P, Randa R (eds) Transnational crime and global security. Praeger Security International. https://books.google.com.ph/books?hl=en&lr=&id=hjFEDwAAQBAJ&oi=fnd&pg=PA301&dq=Weaponization+of+Infectious+Disease+transnational+crime+ra+hamilton&ots=Dxk7qCZmW8&sig=-oMZwp1MCeRBkNVexe60Yh7FCJM&redir_esc=y#v=onepage&q=Weaponization%20of%20Infectious%20Disease%20transnational%20crime%20ra%20hamilton&f=false

International Atomic Energy Agency (2019) Developing a national framework for managing the response to nuclear security events. International Atomic Energy Agency, Vienna. IAEA nuclear security series no. 37-G. https://www-pub.iaea.org/MTCD/Publications/PDF/P1876_web.pdf

Mair M, Mair JS (2003, Summer) A complementary approach to bioterrorism prevention. Nonproliferation Rev 1–11. https://www.nonproliferation.org/wp-content/uploads/npr/102mair.pdf

McLeish C, Nightingale P (2007) Biosecurity, bioterrorism and the governance of science: the increasing convergence of science and security policy. Res Policy 36:1635–1654

OECD (2018) Towards an all-hazards approach to emergency preparedness and response: lessons learnt from non-nuclear events. https://www.oecd-nea.org/upload/docs/application/pdf/2019-12/7308-all-hazards-epr.pdf

OECD (2022) The impact of natural hazards on hazardous installations. https://www.oecd.org/chemicalsafety/chemical-accidents/impact-of-natural-hazards-on-hazardous-installations.pdf
Responding to a CBRNe event: joint operating principles for the emergency services, 1st edn (2016). https://www.jesip.org.uk/uploads/media/pdf/CBRN%20JOPs/JESIP_CBRN_E_JOPS_Document_On.pdf
UN world conference on disaster risk reduction, 2015 March 14–18, Sendai, Japan. United Nations Office for Disaster Risk Reduction, Geneva (2015). http://www.wcdrr.org/uploads/Sendai_Framework_for_Disaster_Risk_Reduction_2015-2030.pdf
Waugh WL (2004) The 'all-hazards' approach must be continued. J Emerg Manag 2(1):11–12
Wright S (2004) Taking biodefense too far. Bull Atomic Sci 60(6):58–66

Chapter 5
Biological/Pandemic Hazards

Emily Ying Yang Chan, Caroline Dubois, and Chi Shing Wong

Abstract For the past decades within the health sector, risk reduction and emergency preparedness against biological hazards have offered important insights to illustrate the usefulness and applicability of the "all hazard approach" to community resilience. The Health Emergency and Disaster Risk Management (Health EDRM) framework (WHO, Health emergency and disaster risk management framework, World Health Organization, Geneva, 2019) is the latest WHO policy and research paradigm adopting the "all hazard" approach. It uses policies and actions in medicine and public health to reduce health-related disaster risks. It also highlights the core building blocks for resilience of individual well-being and population health. By focusing on proactive preventive actions, the framework also argues for a comprehensive multidisciplinary approach in analysing "all hazard" risks within the emergency management cycle, and advocates for an "all hazard" planning and risk reduction approach. Numerous global health emergencies, biological hazards and major crises have seriously affected community health outcomes in a diverse and complex manner (e.g. Ebola, H1N5, SARS, MERS and recently COVID-19 pandemic). These have revealed the technical limitations and practical challenges in "all hazard approach" conceptualization and planning, despite its conceptual merits. There appears to be a lack of relevant research enquiries in evidence-based effectiveness of health risk reduction within scientific literature when comparing "all hazard" with a "specific hazard modular" planning approach. This chapter examines the "all hazard" applications of Health EDRM policy and program implementation for building community resilience. It also discusses current knowledge frontiers and conceptual gaps within the "all hazard" approach that might hamper the planning and development of health systems resilience globally.

E. Y. Y. Chan (✉) · C. Dubois · C. S. Wong
Collaborating Centre for Oxford University and CUHK for Disaster and Medical Humanitarian Response, The Chinese University of Hong Kong, Hong Kong SAR, China
e-mail: emily.chan@cuhk.edu.hk

E. Y. Y. Chan · C. Dubois
GX Foundation, Hong Kong SAR, China

T. Izumi et al. (eds.), *All-Hazards Approach*, Disaster Risk Reduction,
https://doi.org/10.1007/978-981-97-1860-3_5

Keywords Biological hazard · Pandemic · Health emergency · Health Emergency and Disaster Risk Management (Health EDRM) · Health risk reduction · All hazard approach · Top hazard approach

5.1 Introduction

Effective management of disaster health risks for the protection of human well-being has always been a major multidisciplinary challenge at the intersection of health, capacity building and disaster risk reduction (Chan and Shaw 2020). For the past decades, health risk reduction and emergency preparedness in biological hazards have offered important insights to illustrate the usefulness of an "all hazard approach" towards resilience-building in a community. The Health Emergency Disaster Risk Management (Health EDRM) framework was established in 2019 as the World Health Organization (WHO) policy and research paradigm that adopts the all hazard approach (WHO 2019). The framework aims to address and reduce health and disaster risks through policies and actions in medicine and public health. It highlights the core building blocks to ensure resilience of individual well-being and population health.

The Health EDRM paradigm focuses on proactive preventive actions, and argues for the adoption of a comprehensive multidisciplinary approach in analysing "all hazard" risks within the emergency management cycle. It also advocates for an "all hazard" planning and risk reduction approach. This chapter examines the "all hazard" applications of Health EDRM policy making and program implementation within the scope of biological hazards. It also discusses current knowledge frontiers and conceptual gaps within the "all hazard" approach that might hamper the planning and development of health systems resilience globally.

5.2 Biological Hazards Risk Assessment

Biological hazards cover a range of hazards of organic origin that can cause significant loss of life in people, animals and crops, leading to severe environmental and economic losses (UNDRR 2020). While it is common for diseases to circulate within a population, hazard risk increases when the interaction between the disease agent, host and the environment becomes abnormal, such as enhanced virulence or enhanced mode of transmission of a pathogen (Chan 2019). The COVID-19 pandemic was a reminder of the potential for a "biological hazard" to bring major security threat to people's well-being and development. An emerging communicable disease can impact not only health, but also the socio-economic ecology of people globally (Chan and Wong 2022). Within the health-centred, prevention-focused Health EDRM framework, three main approaches are currently available to assess the risks of biological hazards. These are (1) strategic health risk assessment during the pre-event phase, which supports planning for prevention, preparedness, readiness, capacity development, as well as the medium- to longer-term risk monitoring and evaluation; (2) rapid health risk assessment for response intervention planning, based on risk associated with specific or detected events and (3) post-event health risk assessment which emphasizes recovery planning. This last approach can be continuous and aims to update and strengthen the overall risk management system.

In these risk assessment frameworks, biological hazards are classified as "existing" or "emerging" based on available knowledge and existing assessment plans. They are also differentiated into "non-emergency" and "emergency" based on their potential to cause epidemics or pandemics (UNDRR 2020).

Strategies and plans can be developed to prevent and mitigate health risks from biological hazards. Risks within a population may differ as the susceptibility of individuals varies with age, underlying health conditions and socio-demographic factors. Biological hazards include diseases, or other hazards such as insect infestation or plant diseases (WHO 2019). The different organic origins represent a broad variety of pathogens, and thus methods of prevention or treatment, each with unique risk pathways and susceptibility profiles (Table 5.1). To plan for health protection, risk pathways have to be conceptualized (Chan 2019). Undertaking health risk assessment for existing biological hazards with potential to result in an emergency (e.g. malaria, cholera) can be done pre-emptively, given the available knowledge on causes, transmission, treatment protocols and potential impacts. Proactive Health EDRM planning in aspects of prevention, preparedness, readiness, capacity building and surveillance can be undertaken based on strategic health risk assessments. Meanwhile, conducting strategic health risk assessment for emerging biological hazards like COVID-19 is difficult due to the uncertainty in available science and clinical management. Nevertheless, health risk assessment of emerging biological hazards can be achieved as 75% of emerging pathogens are zoonotic diseases. While conducting a strategic risk assessment is not very meaningful nor most urgent once an outbreak emerges, a rapid risk assessment is useful to inform immediate measures of mitigation and prevention of secondary disasters (Chan and Wong 2022).

In line with Health EDRM's principle of multi-disciplinarity, in an emergency arising from a biological hazard, a team of experts from multiple disciplines including epidemiologists, virologists, biotechnologists, health practitioners, surveyors, decision analysts and disaster risk reduction experts should be commissioned by the government or health authorities to conduct a rapid health risk assessment. Such a team will be able to provide crucial information integrated from sources of diverse nature across various relevant disciplines, such as modes and media of disease transmission (air-borne, soil-borne, vector-borne, water-borne, etc.); transmission phases ranging from appearance of disease to cluster and community transmissions; incubation period; transmission rates, severe symptoms, recovery and mortality rate; uneven impact of a biological hazard on different age groups, genders, pregnancy, occupations, disabilities and co-morbidities; reoccurrence odds for recovered patients; pathogen mutation; and availability of detection protocol, treatment, and vaccines. Innovative real-time information and communication technologies (ICTs) like online videotelephony meetings and webinars have greatly enhanced the coordination and

Table 5.1 Subtypes of biological hazard, their mode of transmission and methods for prevention and treatment (Chan 2019)

Subtype (per WHO Health EDRM framework)	Cause or disease transmission	Prevention	Treatment	Examples
Airborne diseases	By breathing in pathogens, such as contact between infected droplets (from coughs, sneezes or spit) and mucous membranes	Physical barrier (e.g. face mask), indoor ventilation, segregation/ quarantine, hygiene and sanitation, vaccination	Disease-specific pharmacological agent to eliminate pathogens, supportive treatment for symptomatic relief	Respiratory diseases (e.g. influenza, coronavirus diseases), tuberculosis, measles, meningitis
Waterborne diseases	By consuming contaminated beverage	Hygiene and sanitation, water treatment or chlorination, vaccination, sewage management	Disease-specific pharmacological agent to eliminate pathogens, supportive treatment for symptomatic relief	Diarrheal diseases (e.g. dysentery, cholera, *E. coli*), hepatitis A, polio
Vector-borne diseases	By prick or bite from a vector (mammal, bird, or insect)	Mosquito nets, protective clothing, pesticide spraying, repellents, environmental cleaning, chemoprophylaxis	Disease-specific pharmacological agent to eliminate pathogens, supportive treatment for symptomatic relief	Diseases transmitted by mosquitos (e.g. malaria, dengue fever, zika), fleas (e.g. bubonic plague), sandflies (e.g. leishmaniasis)
Foodborne outbreaks	By consuming contaminated food	Hygiene and sanitation, food safety practices (e.g. cooking, storage), vaccination	Disease-specific pharmacological agent to eliminate pathogens, supportive treatment for symptomatic relief	Hepatitis A, salmonella, Listeria
Animal-human contact diseases	By direct physical contact with an infected organism or its bodily fluid	Hygiene and sanitation, protective clothing, caution during animal handling (live or slaughtered), culling	Disease-specific pharmacological agent to eliminate pathogens, supportive treatment for symptomatic relief	Influenza (e.g. swine flu, avian flu), Ebola virus disease, venomous bite

(continued)

Table 5.1 (continued)

Subtype (per WHO Health EDRM framework)	Cause or disease transmission	Prevention	Treatment	Examples
Antimicrobial resistant micro-organism diseases	By inadvertent or improper use of antimicrobials which jeopardizes clinical efficiency	Following dosage and therapy guidelines	Use of different medications, combination therapy or different therapy guidelines	Multidrug resistant tuberculosis
Animal diseases	By spreading between livestock, domesticated animals or wildlife through varied transmission pathways	Hygiene and sanitation, vaccination, culling	Disease-specific pharmacological agent to eliminate pathogens, supportive treatment for symptomatic relief, culling	Toxoplasmosis, botulism, foot and mouth disease
Plant diseases	By irrigation, wind, or dissemination by insects, animals or people	Use certified disease-free crop, pruning, disinfecting equipment, burning of infected plants	Chemical treatment (fungicide), burning of infected plants	Coffee rust fungus, head smut fungus (corn, sorghum), bacterial ring rot (potato)
Aeroallergens	By inhalation of contaminated air (natural or chemical), or contact with mucous membranes	Hygiene and sanitation, physical barrier, chemoprophylaxis	Chemical treatment for symptomatic relief, immunotherapy	Conditions caused by pollen, fungal spores, dust mites
Insect infestation	By the pervasive influx, swarming or hatching of insects	Hygiene and sanitation, environmental cleaning, pesticide spraying, proper infrastructure (sealed walls and vents)	Pesticide spraying, trapping, environmental cleaning, infrastructure modification	Conditions caused by locusts, pests, bees

information sharing within this kind of multidisciplinary rapid health risk assessment team at a reduced cost.

5.3 Diverse Biological Hazards: Beyond All Hazard Approach

The idea of the "all hazard" approach is that differing hazards share commonalities that can be targeted centrally across hazard planning. Recently, numerous global health emergencies, biological hazards and major crises have seriously affected community health outcomes in a diverse and complex manner (e.g. Ebola, H1N5, SARS, MERS and recently COVID-19 pandemic). These have revealed the technical limitations and practical challenges in "all hazard approach" conceptualization and planning, despite its conceptual merits. There appears to be a lack of relevant research enquiries in evidence-based effectiveness of health risk reduction within scientific literature when comparing "all hazard" with a "specific hazard modular" planning approach. Some experts argue that given the depth of difference across hazard types, and despite the artificial consolidation of scenarios, the "all hazard" approach fails to deliver optimal results in terms of public preparedness when applied in reality. They propose "top hazards approach" instead, which delineates the prioritization according to local risk indicators. Hazards are then differentially dealt with, where top-ranking hazards are given priority in preparedness and planning activities. The "top hazard" approach retains some of the key benefits of the "all hazard" approach, namely, cost-effectiveness, while offering a more robust framework for achieving better levels of preparedness (Bodas et al. 2020).

The hazard subtypes classified as biological hazards vary in transmission routes, methods of prevention, and known treatments (see Table 5.1). In some instances, hazards are found to have common risk factors. For example, some water-borne diseases and vector-borne diseases transmissions are both attributable to poor water and sanitation management, where drinking water is both at risk of contamination by harmful microbials and at risk of harbouring insect larval habitats. In such instances, an all-hazard integrated risk mitigation and disease management can be effective in disrupting both types of hazards. Interventions may include sewage maintenance or improved placement of water containers (Overgaard et al. 2021).

When considering pathogen characteristics with more specificity, the variances across the subtypes of biological hazards create complexities in all aspects of hazard planning from preparedness, intervention, to response planning. In the case of vector-borne diseases, intervention design must consider not only the interaction between infected individuals, susceptible individuals, and the environment, but also vector management (e.g. insecticide spraying) or mitigation of vector-host interaction (e.g. bed nets). Vectors bring in an additional complexity such that transmission or detection cannot be controlled simply with modification of human behaviour.

Preparedness and interventional planning must consider ecological, entomological and environmental components.

Individual or community behaviour is a core determinant within biohazard risk assessment, which again varies across subtypes and poses challenges to all-hazard approach. Within air-borne diseases themselves, the willingness for vaccination, adherence to personal hygiene or taking personal protective measures are major risk determinant. A serial cross-sectional study conducted during the COVID-19 pandemic in Hong Kong showed that between March and December 2020 the use of face masks increased to 100% under the government-issued mask mandate, and over 90% of respondents continued to practice hand washing. However, avoidance of public transport decreased from 53% to 26%. The study identified socio-demographic determinants for preventive-behaviour uptake, such as age, gender and education. Effectiveness of policy or strategies for risk assessment and risk mitigation will be impacted by pathogenic complexities and social nuances (Chan et al. 2021a). These variations in behaviour will impact the outcomes of health risk assessment and disease risk reduction. Animal-human diseases or plant diseases are not impacted to this extent by human behaviour, but rather by environmental factors or livestock management. Prevention methods and disease risk reduction for such biological hazards are also vastly different (Chan 2019).

Globally, hazard management strategies do not reflect the non-linear transition of biological hazards. For example, outbreaks of a pandemic can emerge in overlapping waves which affect the community at different response or recovery phases. With each new wave that arises before the previous one has ended, the community re-enters a response phase before the initial recovery phase has been completed. The non-linear attributes of biological hazards have rendered the human community vulnerable to protracted crises that persist and increase the community's vulnerability to the cascading risks of multi-hazard that generate complex secondary events and interactions. The inadequacy of available risk management strategies catered for the nuances of biological hazards will undoubtedly challenge the resilience of community health and health systems. To ensure that the global population would be more sufficiently protected against future concurrent, cascading or even interacting hazards, revisiting current risk assessment approaches for biological hazards will be instrumental (Chan et al. 2021b).

Biological hazards also differ from other hazards in the limited availability of knowledge products, well-documented sectoral best practices, cross-sectoral learning, lessons learnt globally and case studies on the management of cascading and simultaneously occurring disasters, among other aspects. Similar to the "all hazard" approach, for the "top hazard" approach to be practical, knowledge products have to align with the terminology and relevant concepts employed in this approach. This will support the administration and sectoral stakeholders to undertake more realistic scenario-planning and develop effective strategies. Effective guidelines and material demonstrating desirable community behaviours can become effective tools for risk communication, raising awareness and enhancing the culture of responsible behaviour and preparedness. Knowledge management and development is a continuous process in general (Shaw et al. 2020). Policymakers, practitioners, researchers and community actors involved in Health EDRM and the development of "top hazard" approach need to be able to access, understand and use the relevant evidence in order to be able to make decisions, develop strategies and take actions that are well-informed, effective and efficient in reducing health risks. This evidence needs to come from reliable research, which has been robustly designed, well conducted and properly reported. This evidence will inevitably also highlight the need for new research to resolve ongoing uncertainties and fill knowledge gaps (WHO 2021).

5.4 Conclusion

The WHO Health EDRM framework is a prevention-focused paradigm that reinforces the "all hazards" approach to disaster risk management. This approach consolidates areas of commonality across various hazard types to develop a best practice for risk assessment, response, communication and intersectoral coordination. While the "all hazard" approach has been effective in top-level planning for emergency response, systems infrastructure and overall conceptualization of global-level risk preparedness, the conceptual gaps arising from the "one size fits all" nature of the all hazards approach renders it not always able to deliver optimum risk preparedness given the specificity and nuances within biological hazards that occur in reality. Experts have proposed a "top hazard" approach as a more robust framework for disaster management.

Biological hazards are complex, have a risk of emerging diseases (never-before seen pathogens), may present in non-linear and cascading nature manner, and have limited knowledge products. Response to biological hazards involves many stakeholders from policymakers to epidemiologists, clinicians and community actors. Given these challenges, all aspects of health protection within the disaster management cycle should consider a comprehensive, multidisciplinary "top hazard" approach which incorporates various risk characteristics. This includes resource prioritization, health risk assessment and building resilient health systems. Strategies for preparedness, management and communication must adopt the appropriate terminology, in line with aspects of disaster risk management beyond health alone. This will create more effective, evidence-based risk assessment systems, as well as response mechanisms that are adaptive in their readiness.

References

Bodas M, Kirsch TD, Peleg K (2020) Top hazards approach—rethinking the appropriateness of the all-hazards approach in disaster risk management. Int J Disaster Risk Reduct 47. https://doi.org/10.1016/j.ijdrr.2020.101559

Chan EYY (2019) Essentials for health protection: four key components. Oxford University Press, Oxford

Chan EYY, Shaw R (2020) Health and disasters: health emergency and disaster risk management in Asia. Springer, Singapore

Chan EYY, Wong CS (2022) Global pandemic and human security: using Health-EDRM risk assessment framework to enhance technology support for DRR. In: Shaw R, Gurtoo A (eds) Global pandemic and human security. Springer, Singapore

Chan EYY, Dubois C, Fong AHY, Shaw R, Chatterjee R, Dabral A, Loyzaga A, Kim Y-K, Hung KKC, Wong CS (2021a) Reflection of challenges and opportunities within the COVID-19 pandemic to include biological hazards into DRR planning. Int J Environ Res Public Health 18:1614. https://doi.org/10.3390/ijerph18041614

Chan EYY, Kim JH, Kwok K-O, Huang Z, Hung KKC, Wong ELY, Lee EKP, Wong SYS (2021b) Population adherence to infection control behaviors during Hong Kong's first and third COVID-19 waves: a serial cross-sectional study. Int J Environ Res Public Health 18:11176. https://doi.org/10.3390/ijerph182111176

Overgaard HJ, Dada N, Lenhart A, Stenström TAB, Alexander N (2021) Integrated disease management: arboviral infections and waterborne diarrhoea. Bull World Health Org 99(8):583–592. https://doi.org/10.2471/BLT.20.269985

Shaw R, Chatterjee R, Dabral A (2020) Integrating biological hazards (including pandemics) into DRR planning: technical advisory document. http://www.ccouc.ox.ac.uk/technical-advisory-document-integrating-biological-hazards-including-pandemics-into-drr-planning. Accessed 21 Dec 2020

UNDRR (2020) Hazard definition and classification review (technical report). United Nations Office for Disaster Risk Reduction, International Science Council, Geneva

WHO (2019) Health emergency and disaster risk management framework. World Health Organization, Geneva

WHO (2021) WHO guidance on research methods for health emergency and disaster risk management. World Health Organization. https://apps.who.int/iris/handle/10665/345591

Chapter 6
Natural Hazards Impacts on Industry and Critical Facility: Natech Risk Management Toward Resilience Building in Indonesia

Farah Mulyasari and Harya Dwi Nugraha

Abstract Natural hazards pose a threat to industrial operations and essential infrastructure. These repercussions can cause the so-called Natech (Natural Hazard Triggering Technological Disasters) accidents, such as emission of poisonous compounds, fire, and explosions, which potentially lead to health damage, environmental contamination, and economic losses. These Natech risks are driven by a variety of variables, including technical concerns related to the features intrinsic in Natech events, as well as governance risk and socioeconomic challenges. This chapter discusses key Natech cases in Indonesia, a country known for its high Georisk leading to high frequency of natural hazard events. These events have severely impacted industries and infrastructures, and thus capture different perceptions and approaches among related stakeholders. We suggest a robust and inclusive risk management, as well as collaboration among stakeholders, are needed to develop policy recommendations to achieve disaster resilience of industrial hazards.

Keywords Natural hazards · Industrial hazards · Natech · Risk management · Indonesia

6.1 Background

The ASEAN Risk Monitor and Catastrophe Management Review (ARMOR 2018) illustrates technological accidents triggered by natural hazards, known as Natech, are often more severe in terms of human mortality, economic loss, and environmental damage than either a natural or technical disaster on its own. Predominately, Natech has been studied with an emphasis on the negative effects of catastrophes on industrial facilities that result in the discharge of hazardous pollutants (hazmat).

F. Mulyasari (✉) · H. D. Nugraha
Center for Sustainable Geoscience and Outreach (CSGO), Universitas Pertamina, Jakarta 12220, Indonesia
e-mail: farah.mulyasari@universitaspertamina.ac.id

T. Izumi et al. (eds.), *All-Hazards Approach*, Disaster Risk Reduction,
https://doi.org/10.1007/978-981-97-1860-3_6

The global chemical industry surpassed $5 trillion in 2017 and is expected to treble by 2030 (UN Environment 2019). Misuri et al. (2020) stated that the manufacturing and use of produced chemicals is spreading globally, with a growing percentage, currently concentrated in emerging and transition economies, many of which may have inadequate regulatory capability.

A document written by the UNDRR and APSTAAG in 2020 illustrated that the Natech occurrences were estimated to account for around 5% of all records recorded in an industrial accident database up to about 20 years ago. On top of that, the manufacturing and consumption of produced chemicals is spreading globally, with a growing percentage presently based in emerging nations and transition economies, many of which may have inadequate regulatory capability. Countries in ASEAN such as Thailand, Vietnam, the Philippines, as well as Indonesia, in relation to chemical industry, presence and expansion has expanded dramatically during the last 25 years, and this trend is projected to continue in the next year. In the same document, UNISDR and APSTAAG also mentioned that these natural hazards causing technological disasters (Natech) are complicated phenomena that were initially investigated at the end of the 1970s and have recently regained relevance. Currently, Natech events are underreported. Furthermore, it is expected that the growing frequency of climate change-related natural disasters may contribute to an increase in the number of Natech cases in the future. In the same report as well, it is stated that there are ten basic principles of Natech risk management that serve as a guidance for the implementation of the Asia–Pacific Regional Framework for Natech Risk Management (UNDRR-APSTAAG 2020). These Ten Basic Principles of Natech Risk Management are as follows:

1. Conduct Multi-hazard, Systematic Risk, and Ecological Impact Assessment

Natech risk management requires a comprehensive risk assessment that considers the susceptibility, capability, and exposure of the community, assets, and environment to different threats. It should entail examining the systemic interdependence of natural and technology risks in a built environment, as well as evaluating the efficacy of current capacity in relation to expected risk scenarios.

2. Enable Coordinated Policy and Planning Among Local Government and Industrial Clusters

Each state bears the primary duty for establishing and implementing risk governance systems, which include institutions, methods, and policies for implementing the current framework at the national, sub-national, and on-site levels. Effective risk management planning for Natech should include active participation of legislative and execution institutions at all levels, clearly laying out time-bound duties and responsibilities across all stakeholders, including private firms and academics. The plan should specify the sources of funds for plan implementation. Where feasible, links to current disaster risk management plans, sustainable development plans, and climate change adaptation plans should be established.

3. Develop and Implement Safety Codes

National and sub-national safety rules and regulations should be designed to address different elements of structural and non-structural safety using an all-hazards approach. These include standards for land use, building, design, materials, and use in order to provide long-term and resilient infrastructure. Codes should include global and regional experiences while being tailored to national and local requirements.

4. Enhance Off-site and On-site Risk Communication

Robust risk management assessment is the first step toward effective risk communication. Risk communication lays the groundwork for informed decision-making and successful risk management by recognizing and shaping stakeholders' risk perceptions. Effective risk communication should take place between the scientific and academic communities and policymakers, as well as between the staff of industrialist and hazardous sites and the local affected community.

5. Strengthen Internal Capacities

Capacity building comprises skill development, resource mapping, and resource scaling up (both human, equipment, and financial). Each hazardous and industrial installation's in-charge is responsible for building on-site mitigation, response, and recovery capabilities commensurate with the nature and magnitude of risks identified. They are also in charge of increasing the off-site community's coping capabilities. Each state is responsible for improving internal capacity at the national and sub-national levels in order to provide timely early warning of Natech and specialized reaction.

6. Build and Reinforce Critical Infrastructure

Critical infrastructure at the national and sub-national levels must be improved and made robust by relevant authorities to ensure the community's continuing social and economic functioning even during disasters. The robust lifeline infrastructure of health, fire, and emergency response contributes to a significant decrease in life loss and injuries, as well as the quick containment of damage and losses to assets.

7. Enhance Response Capacities

The number of lives saved, and the amount of damage and devastation avoided are the litmus tests for response mechanisms. It is based on Natech risk-informed preparedness measures, as well as the readiness and reaction capacity of on-site and off-site responders, including communities. It is the obligation of authorities at the national, sub-national, and local levels to put in place a network of specialized agencies and mutual-aid groups to save lives, control health and other cascading effects, maintain public safety, and address the afflicted community's urgent basic requirements.

8. Promote Business Continuity Management

Each industry and business should assess the Natech risk and implement plans, such as processes, resilient supply chains, and financial mechanisms, to ensure that business, commercial activities, and services continue to operate in the wake of Natech.

9. Plan for Recovery

The sub-national and on-site authorities oversee the safe disposal and management of any hazardous discharged during the Natech. Following this, the impacted community's services, livelihood, and normality are quickly restored. To avoid the formation of future dangers, the recovery process should be linked with sustainable development principles and led by the building back better approach.

10. Foster Multi-stakeholders Partnerships

Natech risk management is an inter- and trans-disciplinary discipline that tries to bring academics, practitioners, and policymakers together at different levels. Depending on the nature of the technical catastrophe, Natech may request trans-boundary collaboration for its handling. It is the role of authorities at all levels to foster such collaborations.

From the aforementioned standpoint, while referring to Indonesia, an ASEAN nation internationally renowned for its high Georisk, resulting in a high frequency of natural hazard occurrences, the ten principles indicated above should be advanced in assessing and controlling its Natech events. The Natech cases are depicted, which date back two decades. Following that, the deficiencies will be examined, and a rigorous and inclusive risk management strategy, as well as stakeholder participation, will be presented to generate policy recommendations for improving disaster resilience of industrial risks.

6.2 Overview of the Natech Case Studies in Indonesia for the Last Two Decades

6.2.1 Case Study: Indian Ocean Tsunami, Indonesia, 2004—Impact on Oil Tanks and Cement Plantation

On 26 December 2004, a 9.1–9.3 magnitude earthquake occurred in offshore western coast of Sumatra (Aceh Province), Indonesia. The earthquake then generated 30 m-tall tsunami wave destroying coastal infrastructures and killing approximately a total of 220,000 people within Indonesia, and 280,000 outside (BBC 2005). The tsunami damaged industries in Aceh with estimated loss of at least US$ 200 million (Detikcom 2005a) and also damaged fuel tanks potentially causing oil spills, and damaged cement plant potentially causing chemical contamination (Detikcom 2005b).

Impact on human life: The tsunami caused the death of at least 130,736 people, with 37,063 went missing and more than 500,000 people were displaced (Meisl et al. 2006). Although damaged fuel tanks and cement plant were thought to be potential source of Natech event, however, there are no available reports on how the damages affected people and the environment (UNDRR-APSTAAG 2020).

Impact on the economy: The local economy, that was mainly supported by fishing and tourism industries, were impacted a lot. Around 66% of the fishing boats and industrial infrastructure in coastal areas have been destroyed by the tsunami, creating major impact on this export-oriented industry (Reliefweb 2005). The extensive media coverage of the tsunami event has created major ripples of tourist arrivals cancelations in other affected areas, such as in Thailand (Rigg et al. 2005). About 45% of cement plantation owned by PT. Semen Andalas Indonesia in Banda Aceh was severely damaged and took more than a year to recover (Detikcom 2005b). The damaged cement plantation had resulted in the shortage of cement supply in the region, at least yearly production of 1.2 million tons could not be produced (Tempo 2005). Despite this major incident, there is no specific data stating chemical contamination caused by the damaged cement plantation.

Impact on the environment: Conversion of mangrove forests into shrimp farms and related environmental destruction worsened the damage by the tsunami, as mangrove could act as filter of waters flowing across the coastal areas (Srinivas and Nakagawa 2008). Furthermore, intrusion of saltwater brought by the tsunami to the subsurface contaminated the groundwater, with about 20% were permanently damaged (Wetlands International 2005). The tsunami was not only extensively damaged and mixed many types of debris in the coastal areas, but the backwash carries these wastes back to the ocean (Srinivas and Nakagawa 2008). Three fuel tanks that were partly filled were displaced from its location, lost the remaining fuel content, and then damaged several houses. Tanks that were nearly full, were not displaced. There was no fire or related Natech event reported. During the first wave, the fuel ship was still in port unloading the fuel to the tanks but managed to escape to open sea when the second wave arrived (Ghobarah et al. 2006). Thus, there was no cascading event following the damaged fuel tanks.

Level of preparedness and measures taken: Aceh was not ready in facing the 2004 tsunami for at least three reasons (UNDRR-APSTAAG 2020): (i) lack of earthquake and tsunami early warning system, (ii) lack of adequate awareness for the risks of tsunami in the community, and (iii) poor mitigation plan and practices for earthquake and tsunami. About fifteen minutes after the earthquake, Hawaii Pacific Tsunami Warning Center (PTWC) informed that a tsunami watch was advised. However, this tsunami warning was not delivered to possible affected countries as they are outside Pacific Ocean (Ghobarah et al. 2006). Thus, no early warning system was utilized to mitigate the risk. People awareness in the main coastal areas of Aceh is relatively low, and the fact that mangroves were converted for industrial purposes worsened the tsunami impact. However, offshore Aceh, in Simeuleu Island, the death toll is surprisingly low; around 5,500 houses were destroyed, and hundreds of people were injured, but only seven deaths were recorded. The reasons for this are (Telford et al. 2006): (i) the coastal ecosystem (coral reef, sea grass, mangrove forests), and (ii) local customs and traditions for mitigating tsunami that has been passed from generation to generation. Damages on infrastructures, that were mainly impacted by debris (e.g., boats, fuel tanks, cars, etc.), were mainly due to inadequate construction quality (Ghobarah et al. 2006).

Lessons learned:

In general, the tsunami has risen the awareness to utilize early warning system in the Indian Ocean. Moreover, the development of tsunami museum is an effort to educate and enhance awareness of the local community on how to be more prepared should the same disaster. The industry is also needed to reinforce the infrastructure quality (i.e., fuel tanks and cement factory), and make contingency plan so that they could reduce impact on people and the environment, and recover quickly, whenever the disaster repeated.

6.2.2 Case Study: Lumpur Sidoarjo, Indonesia, 2006—Earthquake Potentially Induced Drilling Blow out and Mud Volcano Eruption

On 27 May 2006, an earthquake occurred in Yogyakarta with magnitude of 6.3 Mw at about 06:00 local time. Two days later, on 29 May 2006, 250 km from Yogyakarta, a mud volcano called "Lusi" erupted in the city of Sidoarjo (Manga 2007). The eruption covers c.10 km^2 and flowed mud at an average rate of c. 64,000 m^3/day^2 and displaced 40,000 people (Tingay 2010a). Prior to this eruption, within 5–7 h of the earthquake, Banjar Panji-1 exploratory gas well experienced control problems, indicating fluid pressure changes after the earthquake (Manga 2007). This well is suggested to experience an internal blowout and manifested on the surface as a rupture. This surface rupture, c. 150 m from the well, was the origin of the mud volcano eruption (Tingay et al. 2008). On 22 November 2006, subsidence of the mud triggered landslide which then ruptured the East Java Gas Pipeline, resulting in 14 people dead and 13 injured with further 3 people killed in heavy equipment accidents (Tingay 2010b). Despite the competing arguments on the triggers of Lusi (i.e., earthquake vs. drilling), cascading events starting from the earthquake, the drilling internal blow out and mud volcano eruption, may represent a Natech event.

Impact on human live, economic, social, and environment: The eruption of Lusi mud volcano has displaced at least 40,000 people who lived in 12 villages along with their houses, schools, offices, factories, and mosques (Tingay 2010a, b). A total economic loss of $420 M was estimated. Lapindo Brantas (the owner of the well) has agreed to pay Rp 3.8 T based on the area affected by the mud for the affected victims (Raditya 2019). By 2019, compensation for the victims has yet to be completely paid; there are many victims that have not received any compensation yet (Raditya 2019). Other than individual victims, companies whose assets also affected by the mud are still waiting for compensation. As factories owned by those companies cannot operate, many people lost their jobs (Raditya 2019). The continuous flow of the mud has contaminated rivers with heavy metals (The Straits Times 2016).

Level of preparedness and measures taken: Although the Yogyakarta Earthquake in 2006 may have had an indirect impact on the mud volcano, drilling activities were key in initiating it (Mazzini et al. 2007; Davies et al. 2008). It is anticipated that a combination of dangerously drilled windows and a lengthy uncased borehole would have caused a significant surge in the well, making it difficult to control and increasing the risk of blowout. The pressure rise caused by the surge event was sufficient to fracture the formation and establish paths to the surface (Tingay et al. 2008). The unforeseen effect of these dangerous drilling techniques demonstrates the company's inability to deal with the eruption. There are several attempts underway, as preparedness activities, to control the continually streaming mud (Tingay 2010a, b), such as sidetracking of the well, drilling of relief wells to intersect and terminate the open hole of the well, concrete "ball-and-chains" thrown out to the crater but no flow reduction, mud flow diversion and pumping to the adjacent river; even doing prayer and animals scarification they did as measures of preparedness. It turned out that preparedness measures were effective. In 2013, the mudflow began pulsing rather than flowing normally, suggesting that the geological driving force had slowed the mudflow (Amos 2013).

Lessons learned:

Safe drilling practice is needed to be conducted and supervised closely, both by the involved companies and the government. Drilling permit and proposed drilling scenarios need to be evaluated comprehensively, especially in areas with certain geological hazards that can be mitigated prior to drilling.

6.2.3 Case Study: Anak Krakatau Volcanic Eruption and Sunda Strait Tsunami—Potential Affecting Industrial Zones in Cilegon City, Banten Province, Indonesia 2018

On 22 December 2018 at 21:03 WIB (Western Indonesian Time) or 14:03 UTC, Mount Anak Krakatau erupted and damaged nearby seismographic equipment. At 21:27 WIB, the Meteorology Climatology and Geophysics Agency (BMKG) detected a tsunami off the west coast of Banten, although there was no tectonic event. According to existing facts, a landslide occurred from Mount Krakatau as much as 64 hectares which triggered the shaking that led to a tsunami (Ramdhani 2018). National Disaster Mitigation Agency (BNPB) spokesman Sutopo Purwo Nugroho released a statement linking the tsunami to high tides and underwater landslides caused by the eruption of Mount Anak Krakatau (Siswadi 2018).

The eruption of Mount Anak Krakatau Volcano that triggered the Sunda Strait Tsunami in December 2018 has the following impacts:

Impact on human life: Social and economic aspects of the community have been affected by the disaster, especially in Padeglang and Lampung regencies. Based on data from the National Disaster Mitigation Agency (BNPB), it is known that, in Pandeglang Regency there were 429 people died, 1,485 people were injured, 154 people were missing, 16,082 people were displaced (Medistiara 2018a). In South Lampung Regency, 60 people have been found dead, 22 people missing, and 258 people injured (CNN Indonesia 2018). Casualties and damage to the affected are of Pandeglang, Serang district, South Lampung regency, and Tanggamus (Medistiara 2018b). Impact on economic damages: 882 houses, 73 inns, and 60 stalls were damaged. On vehicles, it was recorded that 434 boats and ships were damaged, 24 4-wheeled vehicles were damaged, and 41 2-wheeled vehicles were damaged. In addition, there is 1 damaged dock, and 1 damaged shelter. According to a report from BBC Indonesia, almost all stalls, shops, minimarkets/small supermarkets, including at the tourist beaches of *Anyer*. Piles of rubble are scattered on the edges and corners of the road, mostly by semi-permanent buildings. In addition, in the field of electricity, there are 146 electrical substations that have been turned on. 102 were out, and 20 medium voltage overhead lines have collapsed due to the tsunami (Dwiastono et al. 2018; Irawan 2018).

Level of preparedness and measures taken:

There is none, where available, limited information existed; whether the tsunami event had the implications to technological disaster; reminding that Cilegon City is located at the shore of Anyer Beach (the location of the tsunami) consisted of heavy industrial zones, comprising of 84 industries of medium- to large-scale worldwide industries (Lestari et al. 2018), despite the key policy documents. These key policy documents exist, such as

(1) The Risk Assessment of Cilegon City (developed by BNPB): Cilegon City is vulnerable to natural disasters, such as Anak Krakatau Mountain's volcanic eruptions. In 1883, the Krakatau Mountain eruption resulted in a historic earthquake and tsunami. Cilegon City is also near a possible earthquake and tsunami epicenter in the Indo-Australian and Eurasian plate subduction zones, south of the Sunda Strait. As a result, Cilegon City is at high danger of an 8.5 magnitude earthquake, which might result in a tsunami. In addition to the risk of volcanic eruptions, earthquakes, and tsunamis, Cilegon City is also vulnerable to floods caused by dam breakdowns in the Mancak region of Serang Regency, in addition to coping with sand mining, high rains, and building development. Cilegon City has heavy industrial zones that house medium- to large-scale international companies. Cilegon City is highly vulnerable to natech catastrophes as a result of these circumstances. It is one of 136 cities or districts with significant economic growth and a high natural hazard ranking.

(2) Industrial chemical zones—risk mapping (developed by The Ministry of Health): The Ministry of Health's Health Crisis Centre performed studies on

Chemical Risk Mapping in Cilegon City in 2016. They assessed the Cilegon region to be vulnerable to chemical catastrophes, which may harm more than 47,920 people. According to the research, the chemical accident might have an impact on nearby hospitals, due to cars transporting hazardous chemicals on city streets and highways. As a result, not only industrial zones, but also the road network, are particularly vulnerable to chemical catastrophe hazards. The studies depicted the chemical risk mapping for the industrial area and the path-way route map of hazardous material transported by mobile trucks. The Ministry of Health's analysis, however, has significant limitations. It excludes the impacts and damages caused by the earthquake, as well as the inundation produced by the tsunami. Furthermore, it does not provide a precise evaluation of the societal effect of the chemical accident.

(3) The Recommendation for technological failure in Cilgeon City for disaster risk reduction (developed by BPPT (Badan Pengkajian dan Penerapan Teknologi) or Agency for the Assessment and Application of Technology): The BPPT has created a number of scenarios, one of which involves *Ethylene* leaking in one of the sectors. On the specified location, the consequences are modeled using a computer program called "ALOHA" (Areal Locations of Hazardous Atmospheres), which covers poisonous gas emission and fire dangers. It is indicated that the toxic chemical release during the December 2011 scenario is at a concentration of 7,500 ppm and can spread within 1 h along 275 m from the leakage source at the high hazard level (red zone), along 756 m at the moderate hazard level, and along 1.3 km at the low hazard level. In July 2011, another fire hazard consequences modeling was also performed, and it is suggested that the fire hazard level is at 321 m, which is classified as the red zone with gas concentration at 16.200 ppm (60% lower explosion level (LEL) = flame pocket). Meanwhile, the moderate fire hazards area is within 961 m with gas concentration at 2.700 ppm (60% lower explosion level (LEL) = flame pocket) (10% LEL). According to the findings of the BPPT study, chemical hazardous gas dispersions and fire risks may have a big influence on wide regions and harm a huge number of communities.

(4) The Contingency plan at Cilegon City, facing the possibility of earthquake and tsunami threat (developed by BNPB and the Government of Cilegon City): The BNPB and the Cilegon City administration have created a risk assessment for multi-hazard threats that clearly places Cilegon City in the "HIGH" risk category. Cilegon City government efforts have included the creation of contingency and mitigation strategies for earthquakes and tsunamis. The contingency plan includes the consideration of Natech catastrophes. Cilegon City has established that there are 15 tsunami evacuation spots. The local administration has also begun the process of establishing a tsunami early warning system. A comparison of chemical risk mapping result pictures with tsunami evacuation locations and evacuation routes indicates that some evacuation routes are vulnerable to chemical hazard hazards. These comparisons and interagency talks strengthen the capacity to deal with the Natech crisis in Cilegon.

(5) Exercise and Drills for the 2007 simulation, the ARDEX (ASEAN Regional Disaster Emergency Response Simulation Exercise) 2011, and the ARDEX 2018 (developed by the Government of Cilegon City): Local Development Planning Agency (BAPPEDA) of Cilegon City organized drills for earthquakes, tsunamis, and technological disasters in 2007. The city government, in collaboration with industries, has determined three tier levels of emergency response: (i) Level 1: emergency conditions of which the industry itself can respond, (ii) Level 2: emergency conditions of which the internal company cannot respond and therefore requires help from the industrial zone, and (iii) Level 3: emergency conditions of which the internal company and the industrial zone cannot respond, therefore, requiring support from the activation of the Crisis Centre Cilegon BPBD or from a higher-level Disaster Management Agency (BNPB, National Search and Rescue Agency, and others). Moreover, the ASEAN Regional Disaster Emergency Response Simulation Exercise (ARDEX) in 2011 and 2018; in coordination with the BPBD and BNPB, is a full-scale simulation exercise that seeks to test, practice, review, and evaluate ASEAN's emergency response and disaster management mechanisms. It is conducted on a regular basis, generally every two years, to test regional processes and capabilities. The implementation of ARDEX was originally planned in the year 2011 in Indonesia. However, as major disasters repeatedly occurred in 2010, including the Merapi volcanic eruption and the West Papua flood, the ARDEX was not held in Indonesia. The original concept of the 2011 version is almost similar to that of the 2018 version, because both of them aim at dealing with Natech disaster situations. In the scenario development, it is suggested that the natural disaster involves an earthquake (8.5 M) followed by a tsunami, which thereby causes a technological disaster at an industrial plant. ARDEX 2018 involved a tabletop, command post, and field exercises with a theme, "Strengthening ASEAN's Collective Response Capacity Through National Leadership, Regional Enhancement, and International Support" in collaboration with the BPBD of Cilegon City, the BNPB, and the ASEAN Coordinating Centre for Humanitarian Assistance on disaster management (AHA Centre).

(6) Quantitative fire and explosion risk analysis for LPG storage tank (developed by Lestari and Mahendrodjati 2015): The study suggested that if an LPG storage tank catches fire, the heat radiation will be in the area of 297 m (high heat radiation hazard level—red zone), 421 m (moderate heat radiation hazard level—orange zone) and more than 665 m (low heat radiation hazard level—yellow zone) from the source of storage area The quantitative explosion risk analysis was also conducted for the LPG storage tank and it concluded that the surface explosion would reach 50.4 m (red zone: pressure 8 psi, which can cause major damage to buildings and the tank's surroundings); 860 = m (orange zone: 3.5 psi, which can cause serious injury and significant damage to the surroundings); and 196.0 m (yellow zone: 1 psi, which can damage the glass surroundings). On the other hand, internally the National Oil and Gas Company has conducted a quantitative risk analysis to determine the level of risks for their plants and workers. The study done by Lestari and Mahendrodjati defined that the national

oil and gas industry has been implementing a good process safety management system, industrial safety management system, and other safe work best practices. However, if the worst scenario occurs, the heat radiation can reach the residential area surrounding the oil and gas facility; therefore, further analysis for developing the response system of the facility in collaboration with the surrounding communities needs to be carried out.

Lessons learned: The government city of Cilegon has implemented DRR, shown in the previous section with specific gaps as lessons learned, such as coordination among the stakeholders, such as identifying multi-stakeholder and analyses of their knowledge-level, concerns and issues, and their level of support or resistance; as well sharing the results of their assessment openly among them. To address the lack of coordination in the long run, a participatory Technological Assessment/pTA should be in place as a tool for diverse multisector stakeholders to proceed with discussions about and implementations of Natech DRR and risk management. The pTA promotes direct interaction among members and experts to discuss issues and reach a consensus for mitigation through the provision of information and knowledge of science and technology (Lestari et al. 2018).

6.2.4 Case Study: Extreme Weather—Triggering the Burning Oil Refinery, Balongan Indramayu, West Java, Indonesia 2021

On 29 March 2021, at 00:45 WIB (Western Indonesian Time) in Balongan, Indramayu Regency, the fire at Pertamina Oil Refinery Unit (RU) in Balongan, Indramayu West Java occurred. It is debatable, whether the fire was caused by an explosion triggered by a lightning strike during an extreme weather event, and whether the explosion and fire were due to a leak in the piping network and tanks at the Balongan oil refinery facility (CNN Indonesia 2021a; ANTARA News 2021; Kompas Cyber Media 2021; and BBC Indonesia 2021). According to local residents, on Sunday (28 March 2021) there was a strong smell and on Monday morning there was a loud bang coming from the Balongan Oil Refinery.

The fire at the RU Balongan has the following impacts:

Impact on human life: At least 29 people suffered minor injuries, 6 people were seriously injured, and 932 people evacuated. Another 500 people have also been evacuated from Majakerta Village. A number of other villages were affected, including the villages of Balongan, Rawadalem, Sukareja, Tegalurung, and Sukaurip (all located in Balongan sub-district, Indramayu. 932 refugees as a result of this incident were divided into 3 evacuation points, including the Indramayu Regency Hall, Bumi Patra Sports Center, and the Indramayu Islamic Center Mosque. In addition, as a result of this incident, a resident with a known history of heart disease died from shock

(Yulika 2021). Two villages have been vacated and up to a radius of 2 km will be carried out area evacuation (CNN Indonesia 2021b).

Impact on Economic Damages: Pertamina is calculating losses due to the fire incident and the explosion of four tanks of the Refinery Unit (RU) VI Refinery in Balongan, Indramayu, West Java (Fajar 2021):

(1) Estimated Temporary Losses Reach Hundreds of Billions: The calculation process was being carried out by the management of the state-owned energy sector (Pertamina). The Executive Director of the Institute for Essential Services Reform (IESR) Fabby Tumiwa noted that the losses suffered by Pertamina due to the burning of the Balongan Refinery were quite large; the estimated temporary loss is in the hundreds of billions of Rupiah (IDR currency).
(2) Losses Based on Fuel Distribution to Temporary Stops: IESR's calculation is based on the types of losses experienced by Pertamina. First, the distribution of fuel oil (BBM) is temporarily suspended. The production stop itself was caused by four fuel storage tanks engulfed in flames, such as the T-301, -301G, T301-F, and T-301E tanks. In total there are 72 tanks at the Balongan refinery. Out of 72, there were 4 tanks on burning that contained processed crude oil ready to be sold or distributed.
(3) Losses from Infrastructure to Extinguishing Costs: After the incident, Pertamina had to disburse funds to rebuild four tanks and other supporting equipment, as well as to pay the cost to do the blackout. The firefighting fund is estimated to be quite large. Then, the losses were due to the cessation of refinery operations for the next few days. There were also intangible and tangible costs, due to the investigation process that was carried out for days, the refinery may not run. This means "As long as the refinery is not running, the revenue from the refinery will decrease because there is no production" (Fabby, Executive Director of IESR, April 2021 in Fajar 2021).
(4) Loss of Fuel Import Costs: It is the loss from the difference in the price difference in the cost of importing fuel. Despite, it was safe to reserve fuel supplies, IESR assessed that Pertamina must import to cover 125,000 per barrel per day. This step is in line with the cessation of oil distribution and implies that there will be additional imports costs for the oil. Environmental and linked health and other impacts: The flames are large enough to create a thick puff of smoke that jeopardized human respiration and caused huge amount of carbon emissions (Fajar 2021).

Level of preparedness and measures taken: The fire has been able to be localized inside the bund wall; therefore, around the burning tank, there is an embankment that Pertamina can localize the fire. For extinguishing, Pertamina used foam to the bund wall perimeter and the center of the flame, ensuring that the fire was not spread to other areas. Pertamina has also carried out cooling around the fire location. Several fire engines were dispatched. To prevent collateral fire, Pertamina did the cooling in the surrounding area. There were twofold countermeasures: Firstly, Pertamina controlled oil flows to prevent the oil refinery fires from spreading. The Health, Safety,

Security, and Environmental (HSSE) team of Pertamina's Balongan Refinery together with the Indramayu Regency Fire Department focused on extinguishing fires in the refinery tank area located in Balongan Village. "A normal shutdown was being carried out to control oil flows and prevent the expansion of fires," said PT Refining and Petrochemical Subholding Pertamina International Corporate Secretary *Ifky Sukarya* in an official Pertamina statement, Monday (on 29 March 2021). Pertamina has also prepared a place for evacuation and temporary evacuation for local residents in the Bumi Patra Housing GOR and Indramayu Regency Pendomo. Local residents were asked to remain calm and stay away from the fire location; followingly, the road to the vicinity of the incident was also blocked and ensured that the supply of fuel to the community was not disrupted (Akbar 2021).

Secondly, Pertamina accelerated the handling of the impact of the Balongan Refinery Tank incident by simultaneously socializing and verifying the report data of affected residents. The data was collected through reports of affected residents to neighborhoods' administrators, who are collected to the district. The socialization related to the verification of damage to residents' houses was carried out in Majakerta Village by involving the District Council and the Indramayu Regional Government. Previously, a similar socialization was carried out on 2 April 2021, in Sukaurip, Balongan, and Tegalurung villages, involving various parties, including the Indramayu Regional Government Housing and Settlement Service, Balongan District Council, and local village officials. During the meeting, apart from apologizing to the public, Pertamina also explained the efforts made by the company in overcoming the T-301 tank incident at the Balongan Refinery as well as the impact on the social, environmental, and property sectors. A joint team has been formed by the Indramayu Regional Government, whereas Pertamina team were collecting the data and verifying the losses suffered by residents and thus were responsible for the damage caused by this incident. Socialization and data collection were carried out simultaneously in all affected locations. In the conjunction with that, at locations that have completed socialization and data collection, verification was immediately carried out by a Joint Team consisting of the Technical Team (Housing and Settlement Service), Pertamina, Army, Police, Local Disaster Management Agency at Kuwu sub-district, as well as related neighborhoods' administrators. The entire process was simultaneously and in parallel with the Joint Team; hence the repairs can be carried out immediately, of which they were collecting and verifying the data.

Lessons learned: Pertamina has learnt valuable lessons. To prevent similar incidents from re-occurring, the company had re-identified the security aspects at a number of oil refinery facilities and fuel oil terminals. The company's current priority is to build buffer zones in several of its oil refineries and terminal areas. The existence of a buffer zone will create a sufficient safe distance between the refinery area and the community environment (Meilanova 2021).

An overview of the above Natech Case Studies in Indonesia is illustrated in Table 6.1.

Table 6.1 Overview of Natech case studies in Indonesia

No.	Case study	Cause	Impact	Lesson learned	DRR recommendation
1	Indian Ocean tsunami in Banda Aceh, Indonesia, 2004	Earthquake-generated tsunami 9.1–9.3 magnitude triggered damages to oil tanks and cement plant	**Impact on human live, economic, social, and environment**: • 500,000 displaced, 30,736 casualties, 37,063 missing • Ca. 66% of the fishing boats and industrial infrastructure in coastal areas have been destroyed, creating major impact on the export-oriented industry • Damages to oil tanks and cement plant. Oil tanks: possible oil spill. Cement plant: possible chemical contamination and shortage of cement supply • Estimated loss: USD 200 millions • Destroyed mangrove forest into shrimp farms • Contamination of groundwater due to saltwater intrusion • Displacement of oil tanks	The event rose the importance of risk reduction measures that tailored for (heavy) industries (early warning, risk communication, disaster education, and structural mitigation)	• Early warning mechanisms for different types of stakeholders (community/public, government, industry) • Accessible earthquake-generated tsunami disaster education for different types of stakeholders (community/public, government, industry) • Structural mitigation (strong enforcement for earthquake resistant building codes) with punishing and reward system for industries

(continued)

Table 6.1 (continued)

No.	Case study	Cause	Impact	Lesson learned	DRR recommendation
2	Earthquake potentially induced drilling blow out and mud volcano eruption in Sidoarjo, East Java Province, Indonesia 2006	Earthquake-generated sub-mud volcano eruption of Magnitude 6.3 triggered insufficient drilling capability of the company's operator	**Impact on human live, economic, social, and environment**: • The eruption of mud volcano displaced at least 40,000 people. A total economic loss of USD 420 million was estimated • Drilling blow-out induced by the earthquake, triggered the sub-mud volcano erupted constantly • Factories owned by impacted companies cannot operate, many people lost their jobs • The continuous flow of the mud has contaminated rivers with heavy metals	• Safe drilling practice is needed to be conducted and supervised closely, both by the involved companies and the government • Drilling permit and proposed drilling scenarios need to be evaluated comprehensively, especially in areas with certain geological hazards that can be mitigated prior to drilling	• Risk communication for natural and technological risks along with risk assessment and mitigation for different types of stakeholders (community/public, local government, industries)

(continued)

Table 6.1 (continued)

No.	Case study	Cause	Impact	Lesson learned	DRR recommendation
3	Anak Krakatau volcanic eruption and Sunda Strait tsunami in Cilegon, Banten Province, Indonesia 2018	Volcanic eruption triggering tsunami and potentially damages in industrial zones	**Impact on human live, economic, social, and environment**: • 16,082 displaced, 429 casualties, 22 missing and 258 injured • Hundreds of fishermen's boats and ships damaged • Hundreds of electrical substations outages that potentially impacted to disruptions of the operation of 84 medium- to large-scale worldwide industries in industrial zones of Cilegon City	• Local government coordinated among the stakeholders, such as identifying multi-stakeholder and analyses of their knowledge-level, concerns and issues, and their level of support or resistance; as well sharing the results of their assessment openly among them • Local government coordinated with national government agencies, ministries, a business industries to develop a contingency plan for concerted disaster risk reductions actions to natural and technological risks	• Risk assessment (quantitative and qualitative) that covers • Risk mapping of industrial zones that are susceptible to chemical spill risks, including the analysis of HSSE (health, safety, security, and environmental) • Scenario for mitigating technical failure • Drill exercise
4	Extreme weather (hydro-meteorological disaster) in Balongan-Indramayu, West Java Province, 2021	Extreme weather—triggering the burning oil refinery	**Impact on human live, economic, social, and environment**: • 35 people injured, more than 1500 people evacuated, 2 villages within 2 km radius were vacated • Explosions of 4-unit tanks of refinery cost losses of millions USD to fuel distributions, infrastructure, and fuel import	• Although the safety procedures of HSSE are at hand, to prevent similar incidents from re-occurring, the company had re-identified the security aspects at several oil refinery facilities and fuel oil terminals. The company's current priority is to build buffer zones in several of its oil refineries and terminal areas. The existence of a buffer zone will create a sufficient safe distance between the refinery area and the community environment, which needs strong cooperation and engagement from local government and communities	• More stringent of the enforcement of HSSE procedures and strong risk communications and engagement with local authorities and communities

Responding to the aforementioned Natech cases, the authors identified a silver lining of DRR measures that may work in the long run. Related stakeholders such as local government, business entities, community/public, technical agencies should put their plans in synchronized (in sync); because the efforts are existing, but how to combine to be concerted DRR actions toward Natech Disaster Risks is a challenge that needs to be addressed at the earliest. The section below illustrates such suggestions for DRR and resilience toward Natech risks.

6.3 Inclusive Natech Risk Management in Indonesia for Disaster Resilience

The ten basic principles for Natech Risk Management in Asia–Pacific Regional Framework stretch from the sphere of a multi-hazard and systematic risk assessment, risk communication to business continuity management and partnerships. Seeing from the Natech cases of Indonesia (as examples) dated back two decades ago; there are certain gaps in fulfilling those ten basic principles of Natech Risk Management (Fig. 6.1); these are (i) risk communication, (ii) build and reinforce critical infrastructure, (iii) promote business continuity management, and (iv) fostering multi-stakeholders' partnerships.

In terms of risk communication, off-site and on-site communication prior and during the disasters are at hand; however, its effectiveness should be measured to be able to serve as the basis for informed decision-making. And for these to work, recognizing and shaping stakeholders' opinion and risk perceptions is crucial. In terms building and reinforcing critical infrastructure, Indonesia should not overlook constructing, for example, earthquake resistant buildings. There are no surveys nor data that illustrate how many % of critical/lifeline infrastructure such as health and

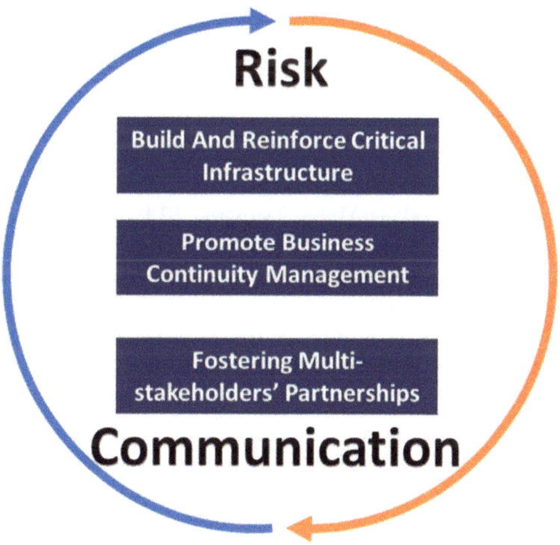

Fig. 6.1 Risk communication framework for Natech in Indonesia

education facilities in Indonesia are "resistant" to/withstand disasters that would tremendously reducing nation's life loss, injuries, assets damages, and economic loss. In terms of promoting business continuity management, nowhere, it cannot be seen a comprehensive business continuity plan for small and medium enterprise partnership program for business entities in bouncing back after they are hit by natural disasters. In terms of fostering multi-stakeholders' partnerships, a coalition to make an ally that is inter-and trans-disciplinary disciplines with concrete and concerted actions cannot be any longer postponed. No single entity/unit/organization can solve/handle/manage the crisis alone; therefore, strengthening the bond and network of local committed and consistent actors in addressing the Natech in their respective job descriptions.

Acknowledgements The authors express their appreciation for parties when carrying out this study.

References

Akbar C (2021) Accelerate refinement fire impact handling, this is Pertamina's strategy (in Indonesian). Tempo.co. Accessed 1 July 2021

Amos J (2013) Mud volcano to stop "by decade's end". BBC News Science and Environment, 13 December 2013

ANTARA News (2021) Polisi sebut ada indikasi kebocoran tangki yang terbakar di Balongan (in Indonesian). ANTARA News Bengkulu. Accessed 2 April 2021

ASEAN Risk Monitor and Disaster Management Review (ARMOR) (2018) AHA Centre

BBC (2005) BBC NEWS—Asia-Pacific, Indonesia quake toll jumps again. Last accessed 1 July 2021

BBC Indonesia (2021) Insiden kebakaran berulang di fasilitas migas, Pertamina didesak benahi 'sistem pengamanan yang tidak andal (in Indonesian). BBC News Indonesia. Accessed 2 April 2021

CNN Indonesia (2018) Korban Tewas Tsunami Selat Sunda Mayoritas Wisatawan (in Indonesian). Peristiwa. Accessed 2 April 2021

CNN Indonesia (2021a) Ahok Ungkap Sebab Kebakaran Kilang Balongan Pertamina (in Indonesian). Ekonomi. Accessed 2 April 2021

CNN Indonesia (2021b) 2 Desa Dikosongkan Imbas Kebakaran Kilang Balongan Pertamina (in Indonesian). Accessed 30 March 2021

Davies RJ, Brumm M, Manga M, Rubiandini R, Swarbrick R, Tingay M (2008) The East Java mud volcano (2006 to present): an earthquake or drilling trigger? Earth Planet Sci Lett 272(3–4):627–638

Detikcom (2005a) Akibat Tsunami, 12 Industri Besar di Aceh Rusak (detik.com). Last accessed 1 July 2021

Detikcom (2005b) Rekonstruksi Semen Andalas Akibat Tsunami Lebih Setahun (detik.com). Last accessed 1 July 2021

Dwiastono R, Henschke R, Wirawan H, Artharini I, Franciska C, Amindoni A, Lesthia, Ginajar G, Susilo M (2018) LIVE: Sunda Strait Tsunami: at least 222 people killed, 843 injured, 28 missing, 556 houses damaged. BBC Indonesia. Accessed 23 Dec 2018

Fajar T (2021) Kerugian Pertamina Imbas Kebakaran Kilang Balongan, Cek 4 Faktanya (in Indonesian). Oke Finance. https://economy.okezone.com/read/2021/04/03/320/2388804/kerugian-pertamina-imbas-kebakaran-kilang-balongan-cek-4-faktanya. Accessed 4 March 2021

Ghobarah A, Saatcioglu M, Nistor I (2006) The impact of the 26 December 2004 earthquake and tsunami on structures and infrastructure. Eng Struct 28(2):312–326

Irawan D (2018) Affected by the Banten-Lampung Tsunami, 102 pln substations are still out. Detikcom. Accessed 23 Dec 2018
Kompas Cyber Media (2021) Kebocoran Tangki Diduga Penyebab Kebakaran Kilang Minyak Pertamina di Balongan (in Indonesian). KOMPAS.com. Accessed 2 April 2021
Lestari F, Mahendrodjati P (2015) Quantitative fire and explosion risk analysis for LPG storage tank (in Indonesian). Internal report
Lestari F, Pelupessy D, Jibiki Y, Putri FA, Yurianto A, Widyaputra G, Maulana S, Maharani CF, Imamura F (2018) Analysis of complexities in Natech disaster risk reduction and management: a case study of Cilegon, Indonesia. J Disaster Res 13(7):1298–1308
Manga M (2007) Did an earthquake trigger the May 2006 eruption of the Lusi mud volcano? EOS Trans Am Geophys Union 88(18):201–201
Mazzini A, Svensen H, Akhmanov GG, Aloisi G, Planke S, Malthe-Sorensen A, Istadi B (2007) Triggering and dynamic evolution of the LUSI mud volcano, Indonesia. Earth Planet Sci Lett 261(3–4):375–388
Medistiara Y (2018a) Update on the number of victims of the Sunda Strait Tsunami: 426 people died, 7,202 were injured, and 23 people were missing. Detikcom. Accessed 25 Dec 2018
Medistiara Y (2018b) Update on the number of victims of the Sunda Strait Tsunami: 429 dead, 154 missing. Detikcom. Accessed 25 Dec 2018
Meilanova DR (2021) This is what Pertamina does so that refinery fires don't happen again (in Indonesian). Bisnis.com. Accessed 1 July 2021
Meisl CS, Safaie S, Elwood KJ, Gupta R, Kowsari R (2006) Housing reconstruction in Northern Sumatra after the December 2004 Great Sumatra earthquake and tsunami. Earthq Spectra 22(3_suppl): 777–802
Misuri A, Landucci G, Cozzani V (2020) Assessment of safety barrier performance in Natech scenarios. Reliab Eng Syst Saf. https://www.sciencedirect.com/science/article/pii/S0951832019306179
Raditya (2019) Sejarah Lumpur Lapindo dan Urusan Ganti Rugi yang Belum Tuntas - Tirto.ID. Last accessed 3 July 2021
Ramdhani J (2018) BMKG's latest update: what happened in Anyer wasn't a tsunami due to an earthquake. Detiknews. Accessed 23 Dec 2018
Reliefweb (2005) Food supply and food security situation in countries affected by the Asia tsunami—Indonesia. Last accessed 1 July 2021
Rigg J, Lisa L, Tan-Mullins M, Grundy-Warr C (2005) The Indian Ocean tsunami: socio-economic impacts in Thailand (PDF). Geogr J 171(4):374–379
Siswadi A (2018) Prima, Erwin, ed. 64 Ha avalanche of Mount Anak Krakatau triggers Sunda Strait tsunami. Tempo. Accessed 24 Dec 2018
Srinivas H, Nakagawa U (2008) Environmental implications for disaster preparedness: lessons learnt from the Indian Ocean Tsunami. J Environ Manag 89(1):4–13
Telford J, Cosgrave J, Houghton (2006) Joint evaluation of the international response to the Indian Ocean tsunami: synthesis report. Tsunami Evaluation Coalition, London
Tempo (2005) Semen Andalas Rusak, Produksi Semen Berkurang 1,2 Juta Ton - Bisnis Tempo.co
The ASEAN Risk Monitor and Catastrophe Management Review (2014)
The Straits Times (2016) Disaster tourism the only, bitter lifeline for survivors after mud volcano swallows Indonesia villages. The Straits Times. May 17, 2016. Accessed 17 Nov 2018
Tingay M (2010a) Anatomy of the Lusi Mud Eruption, East Java. ASEG Extended Abstr 2010(1):1–6
Tingay M (2010b) The Lusi mud eruption of East Java. In: ASEG-PESA conference
Tingay M, Heidbach O, Davies R, Swarbrick R (2008) Triggering of the Lusi mud eruption: earthquake versus drilling initiation. Geology 36(8):639–642
UNDRR-APSTAAG (2020) Asia-Pacific regional framework for NATECH (Natural Hazards Triggering Technological Disasters) risk management. United Nations Office for Disaster Risk Reduction—Asia-Pacific Science, Technology and Academia Advisory Group

United Nations Environment Programme (2019) Global chemicals outlook II-from legacies to innovative solutions: implementing the 2030 agenda for sustainable development-synthesis report

Wetlands International (2005) Tsunami of Aceh and North Sumatra

Yulika NC (2021) Until Tuesday morning, Pertamina Balongan oil refinery fire not off (in Indonesian). Liputan 6. Accessed 30 March 2021

Chapter 7
Cyber Security in the Perspective of Global Risk Landscape

Divya Suresh and Rajib Shaw

Abstract The onset of COVID-19 has accelerated the digitalization process of governments and industries by three to four years. Digitalization aimed toward remote working and creating value for customers has increased dependence on IT networks, making them fallible to cyberattacks. World Economic Forum has listed cyber risk as critical in its "The Global Risks Report 2022". From small and medium-sized firms to federal and large companies, the exposure to cyber risk has increased multifold due to hackers' sophisticated attacks using AI and machine learning. Manufacturing, financial services, and education are some sectors that are seeing an increased momentum of cyberattacks. Amongst the attacks, Ransomware has been the significant cyber threat companies face on their digital supply chain. In addition, attackers are eyeing the gaps caused by emerging technologies, thereby putting the onus on organizations to increase awareness of cyber risks and build defensive mechanisms. In the times ahead, developing sustainable and resilient systems against cyberthreat is essential for achieving safety and security.

Keywords Digitalisation · Cyber risk · Cyber security · Disaster risk · Resilience

7.1 Introduction

Amidst the slow-paced economic recovery post-Covid-19, there are impending risks that daunt all countries' economic, societal, ecological, and technological scenarios in the global landscape. According to the Global Risk Report, 2022 (Mundial et al. 2022), technological risks such as "digital insecurity" and "cyber security failure" are critically short and medium-term threats to the world, while environmental risks fall under the five most critical long-term threats, with "climate action failure," "extreme weather," and "biodiversity loss" ranked as the top three most severe risks. One such

D. Suresh (✉) · R. Shaw
Graduate School of Media and Governance, Keio University, Fujisawa, Japan
e-mail: divi.info@gmail.com

T. Izumi et al. (eds.), *All-Hazards Approach*, Disaster Risk Reduction,
https://doi.org/10.1007/978-981-97-1860-3_7

risk being discussed in this chapter is the Cyber security issue. With increased exposure to cyber risks, everyone has started talking about cybersecurity, from organizations to global leaders. The cybersecurity budgets are expanding. The worldwide estimation for cybersecurity spending by 2025 is US$200 billion (Cybersecurity Founded on 3 Pillars: Technology, Process, and People|Security Magazine n.d.).

In the past few months since Covid-19, the way societies function has changed drastically with rapid digitalization in all sectors of the economy. Covid-19 has made it imperative for people worldwide to be increasingly digitally dependent while exposing them to cyber vulnerabilities. With increasing system insecurity, there is also concern about the increasing digital divide amongst developing nations (Kamel and IGI Global 2010). The digital divide refers to unequal internet access and ICT tools. Worldwide, 52% of women and 42% of men are affected by it. Region-wise analysis shows a much-widened gap, as evident from the data shown by Internet portal world stats as on December 2021: In Africa, only 43.1% of its inhabitants have internet access compared to 88.4% of Europeans and 93.4% of Americans (The Digital Divide—Jan van Dijk—Google Books n.d.). As per the Digital Economy Report 2021, UNCTAD, US and China stand out in terms of capacity to engage in and benefit from the data-driven digital economy. Both countries together account for half of the world's hyper-scale data centers, the highest rate of 5G adoption, 94% of all funding of AI startups in the past five years, 70% of the world's top AI researchers, and almost 90% of the market capitalization of the world's largest digital platforms (United Nations Conference on Trade and Development n.d.).

The traditional digital divide between developed and developing countries remains high regarding internet connectivity, access, and use. This is a persistent challenge for development. With the further evolution and fast progress of the data-driven digital economy, the digital divide is widened by the data-related divide, creating significant power imbalances between countries. So, summarily, 23% of the population in the Least Developed countries have no access to a mobile broadband network; internet use is 90% in developed economies and 20% in LDCs; the average internet speed is eight times that of LDCs in developed economies.

In the context of a country's position along the digital divide, which translates into the level of countries' cybersecurity, cyber threats have outpaced societies' ability to prevent them or respond to them effectively, as there is a dearth of cybersecurity professionals. While preventive measures entail higher costs, it creates cascading physical consequences across societies through disinformation, fraud, and a divergent attitude rather than cooperation among nation-states.

With billions of devices joining IoT every year, cybercriminals are resorting even brighter and bolder moves putting pressure on cybersecurity professionals to run faster. The context of the problem is that with systems and data moving to the cloud, unlike before, securing the network perimeter is not effective as the user can access the system or data from anywhere. Because systems, data, or users are not within the perimeter, cybersecurity strategies need to be reoriented from network centric approach to a user-centric approach. This is the basic idea behind zero-trust architecture. Additionally, security should be considered from various other perspectives, such as security as a code, security as infrastructure, and security as policy.

Cyber regulations are needed to counter cyber security breaches and safeguard data and privacy. It does not entail financial costs, but also the related societal cohesion and mental well-being of people are at stake. Cybersecurity provides prevention and protection against cyber threats and other malicious activities. However, it has been identified that preventive and protective measures are not enough to safeguard systems and related data. In the current scenario, cyber incidents are becoming inevitable, with an increasing number and forms of cyberattacks. These necessitate organizations to build cyber resilience, which is about how the system responds and recovers post a disruption. It mandates the interoperability of cyber security and cyber resilience to provide better security to information systems.

7.2 Cyber Security in Global Risk Landscape

7.2.1 Cyber Issues Vis-A-Vis the Global Risk Landscape

A few cyber security issues are discussed in the below sections: (Mundial et al. 2022).

7.2.2 Convergence of Digital Tools Leading to Cyber Intrusion

Growing dependency on digital systems has increased the dependency on intermediate service providers. The remote working pattern has necessitated the data to be shared with technology-related intermediaries such as cloud service providers, data aggregators, application programming interfaces, etc. This leads to the interconnectedness of multiple technologies such as artificial intelligence, the Internet of Things, the Internet of Robotic Things-enabled devices, edge computing, blockchain, and 5G. Though these capabilities improve efficiency, quality, and productivity, pernicious forms of digital and cyber risks are increased.

7.2.3 Supply Chain Issues

Solar winds Orion attack in late 2020 has shown an enormous scope for contagious exposure, which can break through the defenses of critical cyber security supply chains. The digitalization of physical supply chains is subject to potential contagious threats, as was exposed by a critical security flaw in a widely used software library (Log4j) in December 2021. These growing cyber threats are more significant than societies' ability to prevent and manage them. Moreover, there is a probability for criminal organizations to replicate the older cyberattacks.

7.2.4 Malware and Ransomware Vulnerabilities

With the advent of AI-powered malware, cryptocurrencies, and very few barriers to the entry of participants in the ransomware industry, it has become easy for cyber-criminals, who may even be nontechnical, to use more authoritarian pressure tactics upon vulnerable targets. A few ransomware attacks are encryption of files, data leaks, distributed denial of service attacks, and collection of top executive information for blackmail.

7.2.5 Spyware Technologies

Sophisticated cyber tools are used to create a wave of political and industrial blowback in the form of government sanctions and lawsuits through targeted attacks against journalists and civil rights activists across the globe. Cyber threat actors use deep fake technologies to create social engineering ploys, proliferate disinformation and wreak social havoc, especially during highly volatile times such as that of Covid-19.

7.2.6 Emerging Technological Issues

As society embraces the next version of the internet built upon blockchain technology, one of its manifestations is the metaverse involving a network of 3D virtual spaces enabled by cryptocurrencies and non-fungible tokens (NFT) with unprecedented socio-economic interoperability and immersive virtual reality experiences. Though it may increase the scope and scale of digital commerce, the metaverse can open up more entry points for malware and data breaches. Also, the myriad forms of digital property, such as NFT and digital real estate, further entice cybercrimes. These complex technologies, characterized by decentralization and lack of structures, guardrails, or sophisticated onboarding infrastructure, require the users to navigate the inherent security vulnerabilities.

7.3 Potentials of Cyber Risks

Cyber insecurity cannot be underestimated in the face of the potential risks and challenges it poses in contemporary society. The resultant economic losses are growing substantially across the globe. While the new technologies bring significant opportunities, it simultaneously changes the risks organizations and industries face. The emerging dynamics of cyberspace are increasing organizations' interdependence and mutual reliance on the digital environment, thus creating multiple possible sources of systemic risks. The nature and potential of systemic risk are such that they can propagate through systems and supply chains.

The worldwide accelerated adoption of cyber technologies post-covid-19 has changed the risk equation drastically at all levels ever witnessed before, thus resulting in the following.

7.3.1 Increased and Evolving Threat

Attack automation, speed, and scale of computation and communication have amplified; the range of threat actors has broadened with a stealthy increase of self-evolving malware.

7.3.2 Widened Attack Surface

There is the rapid adoption of new technologies by industries with IoT devices and systems scaling up. High-value infrastructure and resources are increasingly shared. Data generated by devices scale up "big" data sets for AI training with long lifespans of sensitivity. New attack vectors such as insecure technologies, unauthorized access, and outsourcing risks are rising.

7.3.3 Structural Weaknesses

This involves reduced human oversight, reduced visibility of entangled supply chains, and undetected manipulation of algorithms, resulting in the propagation of threats through surface attacks.

7.3.4 Growth in Harm

Critical application of technologies and cyber-physical insecurity have increased the grave consequences. "Monoculture" of service providers and interdependencies between sectors creates harmful effects. This also leads to unintended consequences such as undetected algorithm biases and resource contention in shared infrastructure.

7.3.5 *Impact on Organizations and Economies at All Levels*

With the evolving cyberspace, digital interdependencies increase, multiplying the chances for cyberattack. A greater level of dependency of organizations and advanced economies on digital systems makes them equally vulnerable to cyber-attacks. With digital dependencies accelerating within organizations and economies, the future implication in the event of digital disruption is that the supply chains and different sectors will face risk propagations at levels and speeds that were never witnessed before. Below elaborated are the various levels of the impact caused due to cyber-attacks based on the level of digital dependence (Future Series: Cybersecurity, Emerging Technology and Systemic Risk 2020).

Level 1

Cyberattacks can penetrate successfully through many organizations almost simultaneously due to the pervasiveness of a particular technology or vendors throughout economies. This can lead to incidents and related losses within a short period, affecting multiple supply chains.

Level 2

As organizations increase their reliance on shared technology service providers and supply chains, interdependencies between organizations are ever-growing. This implies that the failure of cyber security in one organization can potentially have a cascading impact on its dependent organizations with systemic consequences.

Level 3

As cyber harms have the potential to cascade through a widening scope of critical functions across industrial and critical infrastructures, cyber security failures could blow systemic catastrophes over economies and societies. Even while the systems are recovered, no alternative providers may be left due to the failure of heterogeneous sectors.

Due to a lack of downstream and upstream transparency in supply chains, many organizations may unknowingly be dependent on components of the cyber ecosystem. As the ecosystem gets more entangled, this dependency will become increasingly unmanageable. With a lack of resilience within the organizations, it may not be possible to account for the aggregation of cyber risk resulting in a growing and hidden cyber-resilience deficit.

7.4 Disaster Risk and Resilience: An Analogy to Cyber Security and Way Towards Cyber Resilience

7.4.1 Risk and Resilience

In recent times, Governments have been put under tremendous pressure to revitalize their national resilience and security after being subjected to a wide variety of threats to public health, cybersecurity, energy networks, and many other interconnected facets of infrastructure, etc. as outcome risk managers and decision-makers of the twenty-first century are put under higher responsibility to devise not only risk management measures but also resilient-driven measures to address unforeseen risk events.

As a result of globalization, the existing systems in various domains, such as social, ecological, engineering, and technical, are complex, interconnected, and interdependent resulting in new challenges. Climate crisis is becoming more prevalent, and in higher frequency, no scientific method still exists to predict the long-term evolution of this crisis precisely. In this case, traditional risk assessment and management proved to be insufficient (Linkov et al. 2013).

With new internal and external threats, resilience thinking and management become the key solution for managers and decision-makers to tackle unpredictable challenges arising from climate extremes, providing an optimal course for protecting critical infrastructures and ensuring systems' sustained functioning. In the (Council 2012) report on "disaster resilience," resilience is defined as the ability of a system to perform four functions concerning adverse events: (i) planning and preparation, (ii) absorption, (iii) recovery, and (iv) adaptation.

The key features of resilience typical for resilience thinking across various fields are critical functions (services), thresholds, recovery through cross-scale (space and time) interactions, memory, and adaptive management (Connelly et al. 2017a, b) The evident difference between risk management to that of resilient management is they may harden the vulnerable system but do not prepare the system enough for low probability yet high consequence risks. Resilience is to be inbuilt into the system additionally, whereby quick recovery and facilitation for adaptation measures are pursued in a crisis (Škanata 2020).

7.4.2 Advantages of Resilience Thinking Over Risk Management

Risk assessment and resilience management are considered to be complementary to each other. They share many common aspects. Risk assessment is the preliminary phase in the resilience management process. This applies to cases of low probability and high consequent events, such as large-scale climate change and cybersecurity threats. The primary difference between risk and resilience management is that resilience thinking is outcome focused. Resilience practitioners think about the affected organization or infrastructure's ability to bounce back from the risk and adapt to new situations. Resilient Management ensures that the hacker breaking cybersecurity or the virulent disease affecting public health or the climate change impacting the environment does not damage the system's functionality and efficiency. The unique perspective of resilience thinking is that it provides consideration to understand the capacity of a system at stake to rebound from massive threats (IRGC 2016).

7.4.3 Cybersecurity

Cybersecurity follows the exact definition of risk assessment and risk management. It helps focus on preventing and protecting computers, programs, data, and networks from malicious activities. Similarly, the goal of cybersecurity follows the same steps as that of formal risk assessment and management procedure as below:

1. Take inventory of Systems and resources
2. Identify potential weaknesses and threats
3. Determine the risk impact
4. Develop and set cybersecurity controls
5. Evaluate the effectiveness of security controls.

Despite pursuing such disciplined practices, cyber threats and risks have only drastically increased and turned out from a possibility to an inevitability. This implies that cybersecurity is not enough anymore (Sharkov 2016).

7.4.4 Cyber Resilience

The above scenario has mandated that nations, Organizations, leaders, and risk managers shift from a cybersecurity perspective to cyber-resilient thinking. Together Cybersecurity and Cyber resilience open a new perspective and are the solution for the way forward. Cybersecurity is the precursor to cyber resilience. Cybersecurity can be standalone. On the other hand, cyber resilience cannot exist without cyber security. Definition of cyber resilience from (Building Cyber Resilience. At a Glance 2018) cyber resilience is an organization's ability to limit the impact of cyber disruptions, maintain critical functions, and rapidly re-establish normal operations following a cyber incident.

Referring to the Cyber resilience framework, as proposed by (Netmarcom n.d.) and (Framework for Improving Critical Infrastructure Cybersecurity, Version 1.1 2018), is based on five pillars as below:

1. Identifying capabilities and vulnerabilities
2. Protecting and securing vital infrastructure
3. Detecting security threats as soon as possible
4. Responding to breaches properly
5. Recovering quickly and efficiently with little downtime as possible.

Moreover, a resilient cyber strategy should encompass the five aspects of pillars such as identity, protect, detect, respond and recover. Utilizing the framework, organizations can improve their cybersecurity risk management (Weil and Murugesan 2020).

7.5 Emerging Cyber Challenges and Way Forward

As organizations are interdependent in cyberspace, it is impossible to encounter digital threats with their capabilities. It is imperative for them to jointly address the systemic risks and thereby secure the digital ecosystem. In this process, numerous vital challenges need to be handled through collective action (Bridging_Cybersecurity_30.06 n.d.).

7.5.1 Bridging Cyber Security Skill Gap Through Education and Training

Alongside the already existing dearth of the skilled cyber security workforce, the emergence of more and more new technologies is widening the skill gap. Adding to the workforce challenge is that people working in this sphere are wearing many other hats resulting in less scope for specialization and innovation.

Accelerated education and training on cybersecurity are required to fortify our walls against cyber criminals and fill the cyber security skill gap. It is recommended to lower standards at the entry-level and hire enthusiasts with an appetite to learn who can be continuously trained and educated in this domain. The workforce at all hierarchical levels must be cyber literate to make meaningful investments toward building cyber resilience. A new generation of operational cyber security capabilities should be developed to overcome the fast-growing cyber risk landscape. Adequate availability of expertise should be ensured to operate in a blended operational technology or information technology environment.

7.5.2 Combatting the Quantum Security Risk

The enterprise leaders will have to be educated and equipped to:

- Assess the materiality of the quantum threat to cryptography, to the organization's assets, ensure the adequate management of cryptographic estate for transition to quantum-safe cryptography within a suitable timeline
- Consider the other security risk controls that the attacker with quantum computing power might weaken
- Decide on adopting quantum computing into the business model, including using quantum-based mitigations of potential security risks.

At the national level:

- Governments and industry should work together on their investment strategy for the development of sovereign quantum capabilities
- Equitable access to quantum technology should be ensured to avoid widening inequalities in security and industrial capabilities between nations
- Alignment between technological standardization and governance efforts worldwide should be provided for smooth international collaboration and the operation of international organizations.

7.5.3 Fragmented Technical and Policy Approaches—A Case for Integrated Cyber Standards and 'License to Participate'

Technical standards and cyber governance principles are divided into jurisdictional and sectoral siloes. Geopolitical divergences and protectionist stances further create divergence risks at the public policy level. Thus, it could only realize the potential value of a secure global digital ecosystem, leading to incompatible security-compliance requirements and costly sub-optimal security systems.

The efforts of intergovernmental organizations towards developing common principles for digital security, international policy recommendations, and international norms should result in integrated standards for the emerging digital era. There should also be a process for identifying emerging and material inconsistencies in regulatory and policy approaches. This will help to resolve inconsistencies in the hyper-connected ecosystem.

The concept of "license to participate" should be well established so that a basic level of security assurance and accountability would be consistently exhibited in the market by the players participating in the cyber ecosystem. Though this concept is well established for organizations defined as "critical national infrastructure" (CNI), which require a license to operate and are subject to regulatory requirements, some infrastructures beyond the CNI obligations depend on shared underpinning digital infrastructure. It is necessary to comply with safety and availability standards of international interoperability.

7.5.4 Uncover Technology Gaps and Enhance Security Capabilities

The present technology innovations need to consider security as its integral component. There is very little or no consideration for malicious threats, resulting in insufficient security functionality, followed by costly retrofit. Existing capabilities and technologies are not sufficient to address emerging cyber threats. New approaches are needed to keep up with the pace and scale of joint operations. The level of automation within cyber defense capabilities should be increased to ensure that the cyber security tools can interoperate effectively (Future Series: Cybersecurity, Emerging Technology, and Systemic Risk 2020).

The risk due to interdependence arising from ubiquitous connectivity can be countered with 5G technology. This will effectively enable faster and more reliable mobile communications for many devices, promising agile and tailored networks with unparalleled computational power analytics supported by the cloud.

Faster and more dynamic AI-enabled defenses should be evolved to counter the pace and sharpened predictive capabilities of AI-enabled cyber criminals and other malign actors. At the same time, it is necessary to consider reducing the potential harms of an emerging arms race, as the attackers will learn from defenders' research to improve their capabilities. Methods of incentivizing equitable and affordable access to existing and emerging defensive technologies should be identified by the government and enterprise leaders. Best cyber security practice guidance should be created towards automated detection, responses, and investigations tailored to the emerging AI-enabled threat landscape and defensive opportunities.

AI security principles should be developed to ensure the following:

I. **Secure design**: AI systems designs should enable those responsible for algorithmic decisions to verify their integrity against adverse manipulation and disruptive techniques.
II. **Life cycle management**: Version control of dynamic AI models is necessary through rigorous vetting of algorithms as new data is ingested.
III. **Incident management**: Whenever the algorithms are manipulated, those responsible for their output should be able to detect the manipulation, reduce its harmful consequences and recover to a state of algorithmic integrity.

7.5.5 Bridging the Accountability and Transparency Gap

Complexities and entanglement of the digital ecosystem are worsening the lack of clarity about accountability and liability for assuring end-to-end services and shared critical infrastructures.

Suppose customers are assured of making the right risk-management decisions. In that case, creating a base level of trust within the supply chain, with transparency in security functions, is essential. Though little sectoral and country-specific progress is being made in this context, there is still a greater need for clear policies and regulations that mutually and internationally support each other. Whenever any transfer of liability occurs, there should be clarity so that those responsible for delivering cyber security functionality should do so with their accountability. Governmental and quasi-governmental actors should realize their potential legal and ethical obligations to protect the critical national infrastructure, reduce the reasonably foreseeable cyber risk-related vulnerabilities and ensure appropriate remedies should the harm occur.

7.5.6 *Insurance Challenges*

Emerging technologies and systemic risks make it difficult for insurers to assess the organization's risk exposure, underwrite aggregated losses and understand the total risk exposure of an insurance portfolio. Moreover, insurance against cyber risks is precarious as insurers themselves are subject to retaliatory attacks for attempting to curb ransomware payments.

7.5.7 *Other Challenges*

- Essential systems of the economy, such as public utilities, health care systems, data-rich companies, GPS, and air traffic control systems, are vulnerable to scathing cyberattacks, thus disrupting the functioning of society at large.
- Cyber espionage attacks may target intellectual property, resulting in developmental and reputational costs to private and public sector organizations.
- Deep fake technologies can be used to manipulate election or political outcomes, thus creating mistrust between societies, businesses, and the government.
- The individuals subject to digital inequality and insecurity are more vulnerable than the wealthier ones.
- Governments at all levels are burdened with mounting pressures to strengthen and secure the digital landscape by responsibly managing the cyber challenges.

As developed countries are way ahead towards multifaceted cyber security approaches, developing countries are framing their digital infrastructure policies amidst their internal issues. On the other end, underdeveloped countries are still struggling to comprehend being at the foot of digital transformation. Developing nations are trying to imitate the robust cybersecurity approaches of developed nations, whose capabilities can be attributed to their investment in human resources, starting with proper education and training. The cyber security gap between developed and developing countries can be bridged by nurturing knowledge, upskilling, and building competencies (Cyber Security Policy in Developing Countries—Secon n.d.).

7.6 Conclusions

Cyber risks entail financial costs, and the related societal cohesion and mental well-being of people are at stake. Leaders have to futuristically plan and invest in cybersecurity research and development to tackle overwhelming risks in the information society. They must also make modifications at the policy level to address crises. A Nation or an organization‘s capability to respond to risk situations depends not only on its effective planning but also on its effectiveness with preparations, conducting trials, and training of its members, the latter part of which is very much neglected. Proactive management strategies should be developed to advance resilience rather than practicing reactive measures. Proactive management strategies imply conducting an attack on your system and identifying the weakness, thereby fixing the drawbacks, termed regular penetration testing. Some comprehensive measures that leaders can take to build resilience are mentioned below.

Risk-informed strategies

According to UNDRR 2019 guidelines, countries’ planning and decision-making for disaster risk reduction activities and investment strategies should integrate technological risks, notably cyber security threats.

Integration of cyber risk in national risk assessment

In line with the Sendai Framework, governments across the globe should consider adopting an all-hazards approach as one single plan covering all possible threats and risks, including hybrid threat scenarios touching both physical and digital spaces. Through trustworthy digital technologies, cyber resilience could be achieved with persistent and intentional trust-building initiatives among global cyberspace players.

Stress test capability to withstand hybrid and cascading risk scenarios

To better understand current capacities and improvements, stress testing tools developed by UNDRR are required to identify complex scenarios, including cyber-related threats. Such stress testing mechanisms will reduce the potential impact of impending dangers and risks and thus ensure business continuity through better understanding and strengthening of the abilities of various sectors and services.

Information sharing

As a part of collaborative operational security measures to address the growing systemic risks, organizations should share information about the threats targeting them and the indicators of compromise and responses among a distributed set of potential cyber actors. This will improve the organizations’ collective defense capabilities and raise the cost of cybercrime for attackers. The information about AI attacks and defense algorithms can be shared and communicated by codified methods. Timely sharing of information will address high-speed, large-scale threats. Organizations should cooperate and share their best practices that can be replicated across industries and economies.

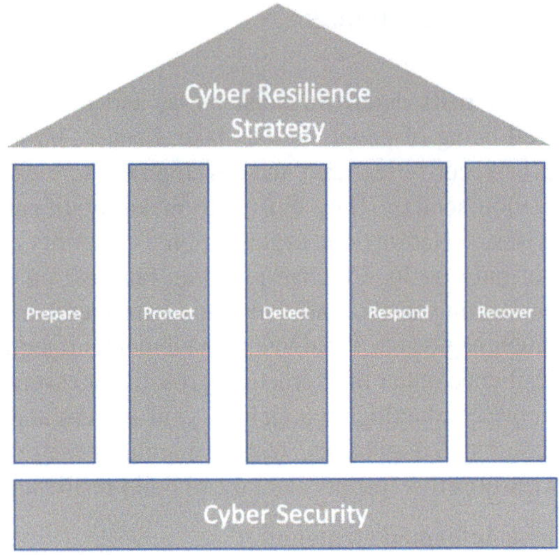

Fig. 7.1 Cyber resilience strategy. *Source* Building Cyber Resilience. At a Glance (2018)

In the scenario of security and world order, during every era, societies have witnessed that nations have resorted to technological advancements to keep themselves adept in preserving their national boundaries secure from prevailing threats. Every time with an advantage of added power (firearms, cannons, nuclear), nations attempt to assess what losses and risks a potential victory on an opponent could bring. Envisioning the possibilities and outcomes, different countries set their objectives and strategies for resilience building in the event of an attack. The related technologies for various types of risks, such as nuclear, cyber, biological, industrial, etc., exist and will individually have a significant part to play in strategy. However, with enormous possibilities with digital systems and AI capabilities and added extraordinary complexities, devising strategy has become more elusive in the current era. This is because, in the case of nuclear or conventional weapons, which exist in the physical space, it was possible to roughly estimate the opponents' capabilities which entirely becomes impossible in the cyber realm, which operates in opacity. The users do not disclose their existence nor acknowledge their capabilities. Also, incorporating such capabilities into the strategy and international equilibrium is challenging because government alone cannot have the required technological preeminence. Entrusting it to one actor will result in preconceptions. The task will mandate the participation of various actors and institutions ranging from the national government, individual inventors, entrepreneurs, research institutes, and startups while considering the strategic implications. This will help see the bigger picture. Not all the actors at once will reach a consensus as their missions will not collide with the national objective. Mutual education and orientation will help bridge the gap in arriving at a common conceptual framework for an all-encompassing strategy (The Age of AI: And Our Human Future—Henry A Kissinger, Eric Schmidt, Daniel Huttenlocher—Google Books n.d.).

Acknowledgements The first author is thankful to Japan Science and Technology (JST) for the scholarship supporting Ph.D. research.

References

Agriculture Industry on Alert After String of Cyber Attacks. https://www.govtech.com/security/agriculture-industry-on-alert-after-string-of-cyber-attacks. Accessed 5 April 2023

Bridging_Cybersecurity_30.06. (n.d.)

Building Cyber Resilience. At a Glance (2018) www.e-library.ca

Connelly EB, Allen CR, Hatfield K, Palma-Oliveira JM, Woods DD, Linkov I (2017a) Features of resilience. Environ Syst Decisions 37(1):46–50. https://doi.org/10.1007/s10669-017-9634-9

Connelly EB, Allen CR, Hatfield K, Palma-Oliveira JM, Woods DD, Linkov I (2017b) Features of resilience. Environ Syst Dec 37(1):46–50. https://doi.org/10.1007/S10669-017-9634-9/TABLES/1

Council NR (2012) Disaster resilience: a national imperative. Disaster resilience: a national imperative, pp 1–244. https://doi.org/10.17226/13457

COVID-19 has speeded the adoption of digital technologies by several years. https://www.mckinsey.com/business-functions/strategy-and-corporate-finance/our-insights/how-covid-19-has-pushed-companies-over-the-technology-tipping-point-and-transformed-business-forever. Accessed 5 April 2023

Cyber Criminal Actors Targeting the Food and Agriculture Sector with Ransomware Attacks, September 1, 2021. https://www.aha.org/system/files/media/file/2021/09/fbi-tlp-white-pin-cyber-criminal-actors-targeting-food-agriculture-sector-ansomware-attacks-9-1-21.pdf. Accessed 5 April 2023

Cyber security policy in developing countries—Secon (n.d.) Retrieved 18 Nov 2022, from https://seconcyber.com/cyber-security-developing-countries/

Cyber security: global food supply chain at risk from malicious hackers. https://www.bbc.com/news/science-environment-61336659. Accessed 5 April 2023

Cybersecurity comes of age as industries transform. https://www.pwc.com/us/en/services/consulting/cybersecurity-risk-regulatory/library/global-digital-trust-insights/sector-analysis.html. Accessed 5 April 2023

Cybersecurity founded on 3 pillars: technology, process, and people|Security Magazine (n.d.) Retrieved 5 April 2023, from https://www.securitymagazine.com/articles/98489-cybersecurity-founded-on-3-pillars-technology-process-and-people

Cybersecurity founded on 3 pillars: technology, process, and people. https://www.securitymagazine.com/articles/98489-cybersecurity-founded-on-3-pillars-technology-process-and-people. Accessed 5 April 2023

Cybersecurity trends: looking over the horizons. https://www.mckinsey.com/business-functions/risk-and-resilience/our-insights/cybersecurity/cybersecurity-trends-looking-over-the-horizon. Accessed 5 April 2023

Framework for Improving Critical Infrastructure Cybersecurity, Version 1.1 (2018) https://doi.org/10.6028/NIST.CSWP.04162018

Future series: cybersecurity, emerging technology and systemic risk (2020)

Growing digital dependency will intensify cyberthreats. https://www3.weforum.org/docs/WEF_The_Global_Risks_Report_2022.pdf. Accessed 5 April 2023

IRGC (2016) IRGC resilience resource guide (Volume 1) (n.d.)

Kamel S, IGI Global (2010) E-strategies for technological diffusion and adoption : national ICT approaches for socioeconomic development. Information Science Reference

Linkov I, Eisenberg DA, Bates ME, Chang D, Convertino M, Allen JH, Flynn SE, Seager TP (2013) Measurable resilience for actionable policy. In: Environmental science and technology, vol 47, no 18, pp 10108–10110. https://doi.org/10.1021/es403443n

Mundial FE, Marsh & McLennan, SK Group, Zurich Insurance Group (2022) The global risks report 2022. World Economic Forum

Netmarcom S (n.d.) Symantec white paper—the cyber resilience blueprint: a new perspective on security

Preparing for a Black Swan Cyberattack. https://hbr.org/2016/09/preparing-for-a-black-swan-cyberattack. Accessed 5 April 2023

Ransomware cemented its position as the most prominent cybersecurity threat. https://www.pwc.com/gx/en/issues/cybersecurity/cyber-threat-intelligence/cyber-year-in-retrospect/yir-cyber-threats-report-download.pdf. Accessed 5 April 2023

Sharkov G (2016) From cybersecurity to collaborative resiliency. In: SafeConfig 2016—proceedings of the 2016 ACM workshop on automated decision making for active cyber defense, co-located with CCS 2016, pp 3–9. https://doi.org/10.1145/2994475.2994484

Škanata D (2020) Special issue on cyber-security of critical infrastructure improving cyber security with resilience. Ann Disaster Risk Sci 3(1). https://ojs.vvg.hr/index.php/adrs/article/view/43/38

Supply chain reinvention needed. https://www.weforum.org/agenda/2022/05/reinvent-supply-chains-pandemic-ukraine/. Accessed 5 April 2023

The Age of AI: And Our Human Future—Henry A Kissinger, Eric Schmidt, Daniel Huttenlocher—Google Books. (n.d.). Retrieved 5 April 2023, from https://books.google.co.jp/books/about/The_Age_of_AI.html?id=kKckEAAAQBAJ&redir_esc=y

The Digital Divide - Jan van Dijk - Google Books (n.d.) Retrieved 18 Nov 2022, from https://books.google.co.jp/books?hl=en&lr=&id=6DvKDwAAQBAJ&oi=fnd&pg=PT7&dq=digital+divide&ots=6ApPyFWisC&sig=MQxvcxYMZ0LVqpy9QuAaUzb-aQQ&redir_esc=y#v=onepage&q=digital%20divide&f=false

Trend No. 3: Digital supply chain risk. https://www.gartner.com/en/articles/7-top-trends-in-cybersecurity-for-2022. Accessed 5 April 2023

United Nations Conference on Trade and Development (n.d.) Digital economy report 2021: cross-border data flows and development: for whom the data flow

WEF_Global_Cybersecurity_Outlook_2022. https://www3.weforum.org/docs/WEF_Global_Cybersecurity_Outlook_2022.pdf. Accessed 5 April 2023

Weil T, Murugesan S (2020) IT risk and resilience-cybersecurity response to COVID-19. In: IT Professional, vol 22, no 3, pp 4–10. IEEE Computer Society. https://doi.org/10.1109/MITP.2020.2988330

Chapter 8
Natural Hazards Triggering Technological Accidents (NATECH) and Future Perspectives

Kumiko Fujita, Takako Izumi, Miwa Abe, and Rajib Shaw

Abstract Natural Hazards Triggering Technological (Natech) Accidents are relatively rare. However, if they occur, the scale can be catastrophic, and the impact on the environment and economy can be enormous. In this research, the recognition and research progress of Natech accidents worldwide are reviewed, and regional differences in Europe, Asia, and the United States of America (the US) are compared. Europe is the leading region in Natech accidents research. Less research has been conducted in Asia and the US. Natech accidents have been recognized in Europe, and research has been conducted steadily. In addition, European countries have implemented common safety management measures and policies against Natech accidents. In contrast, each Asian country takes separate measures, and recognition and research progress vary. The US' approach toward Natech accidents differs from that of Europe and Asia; that is, it focuses on responding to Natech accidents rather than preventing them. After reviewing these differences, future perspectives of the Natech approach are discussed.

Keywords Natech · Europe · Asia · The US

K. Fujita (✉) · T. Izumi · M. Abe · R. Shaw
International Consortium On Landslides, Kyoto, Japan
e-mail: kumiko.fujita@gmail.com

T. Izumi
Graduate School of Internatoinal Cultural Studies, Tohoku University, Sendai, Japan

M. Abe
School of Humanity and Sciences, Tokai University, Kumamoto, Japan

R. Shaw
Graduate School of Media and Governance, Keio University, Fujisawa, Japan

T. Izumi et al. (eds.), *All-Hazards Approach*, Disaster Risk Reduction,
https://doi.org/10.1007/978-981-97-1860-3_8

8.1 Introduction

Natural Hazard Triggering Technological (Natech) Accidents are industrial accidents caused by natural events, such as hurricanes, floods, earthquakes, tsunamis, etc. There are several definitions given by different organizations as follows:

Organisation for Economic Cooperation and Development (OECD):

Natech is a chemical accident triggered by natural hazards/risks from natural hazards at hazardous installations (OECD 2022).

European Commission (EC):

Technological accidents triggered by a natural hazard or disaster which result in consequences involving hazardous substances (e.g., fire, explosion, toxic release) are commonly referred to as Natech accidents (eNatech 2022).

Natech accidents have almost never occurred in countries with few natural hazards, as the possibility of natural hazard-induced disasters is rare, such as in Middle Eastern countries. In addition, even though countries have a high frequency of natural-induced disasters, if the area is not industrialized, the possibility of Natech is few. Thus, Natech accidents can be observed in industrialized countries with natural hazards and natural hazard-induced disasters, such as the US, Europe, and some Asian countries. Even in these countries, Natech accidents are unusual; however, their consequences can be catastrophic. Although the frequency of Natech accidents is currently low, there is a possibility that the number of occurrences will increase, and the damage will also become considerable due to extreme natural hazards caused by climate change and the acceleration of industrialization. Therefore, it is necessary to review the trends in Natech research and discuss future perspectives of the Natech approach.

In this chapter, catastrophic Natech accidents, which make people recognize that there are natural hazards that trigger technological accidents, are introduced. The research progress and strategies of Natech in each region are then reviewed. Based on the results, this chapter aims to summarize the trends in Natech research worldwide, and the future perspective of the Natech Approach is summarized.

Some catastrophic Natech events received attention and gave impact on the people's thinking way and behavior. One of them is the earthquake of August 17, 1999 at Kocaeli, Turkey. It was the largest and the most destructive earthquake in Turkey since the 1939 Erzincan earthquake (Toksoz et al. 1999), and many Natechs occurred due to the earthquake. Although it was clear that the earthquake-induced multi-hazardous material release in Turkey was a Natech accident, it was not understood as Natech yet at that time as the word Natech had not been well-known. This accident made people aware of the accident to be called Natech. The major two Natech events were massive fire at the TUPRAS Izmit petroleum refinery and the hazardous acrylonitrile spill at the AKSA acrylic fiber production plant. The fires burned for several days (Scawthorn et al. 2005). It highlighted problems in the consideration of Natechs in emergency planning, response to industrial emergencies during the natural disasters and information to the public during and following the incidents (Girgin 2011).

The 2002 flood in Europe also affected the people's thinking way. It was the most destructive event in terms of economic losses from 1999 to 2009. In August 2002, a week of intense rainfall produced floods. It reached the Czech Republic, Italy, Spain, Austria, Germany, Slovakia, Hungary, Romania, Bulgaria, Croatia, and Ukraine (Svetlana et al. 2015). It motivated a relatively large number of people to take individual precautionary measures (Kreibich et al. 2005).

Hurricanes Katrina and Rita caused huge losses to the oil and gas industry, and had economic effect all around the world. In addition, the aftermath caused severe long-term damage on environment. Clean up the hazmat releases costs a lot. This is a particularly important issue not only in the Gulf of Mexico but also in all Gulf Coasts, where hurricanes are frequent, industrial facilities are numerous, and high population is dependent on coastal and estuarine resources. Hurricanes Katrina and Rita gave opportunity for governments and local people to think how to better prevent future releases and be better prepared for future storms (Cruz and Krausmann 2009, Picou 2009).

Thus, the Natech accidents caused by the earthquake and tsunami in Japan demonstrate the risk of cascading effects, such as multiple and simultaneous releases of hazardous materials over wide areas (Krausmann and Cruz 2013). Japan is considered to be a country of well-prepared for natural disasters. However, it dramatically highlighted the vulnerability of hazardous installations to the effects of natural hazards (Krausmann and Cruz 2013).

Table 8.1 lists four selected Natech accident examples. These disasters gave researchers the opportunity to pay attention to Natech accidents because of the impacts. Natech research was first focused on earthquakes as the main triggering event, and later the focus shifted to hydrometeorological hazards such as floods, multi-hazard studies and cross cutting issues (Cruz and Suarez-Paba 2019).

8.2 Natech Research

The Natech concept was first introduced in the US by Showalter and Myers (1992) while Natech accidents occurred even before the term is recognized. After the phenomenon was named "Natech," it garnered a lot of attention. Figure 8.1 shows the trend in several English research papers with the word "Natech" in the title, based on a Google Scholar search in January 2022. The term "Natech" has been widely used since the 2000s.

This increase in number is a useful barometer for the increasing recognition of Natech. Since the Internet was not common before the 2000s, there were few digitalized research papers. Therefore, the reason the number of research papers on the Natech accident in Turkey in 1999 were few might be because not only was Natech not well recognized, but the internet system was also not common, and digitalized papers might have been rare.

Major Natech research has been conducted in European countries. Before 2010, most research focused on case studies, especially on events occurring in Turkey,

Table 8.1 The past Natech accidents

	Country	Natural hazard	Technological accident	Impact/lessons learned
1999	Turkey	Kocaeli earthquake	The affected area was a densely populated area in one of the most industrialized regions in Turkey. The earthquake damaged more than 350 industrial firms, and multiple hazardous materials, such as naphtha, ammonia, and toxic acrylonitrile, were released into the air, water, and soil	Realized the potential danger of a Natech accidents and affected emergency planning for responding Natech accidents
2002	Europe	Floods	400 kg of chlorine gas was released because of the flood caused by the Spolona Chemical Works company in the Czech Republic, which is on the River Elbe that swept across Europe	Motivated a large number of people to take private precautionary measures
2005	USA	Hurricanes Katrina and Rita induced flood	Two hurricanes impacted the Gulf of Mexico on August 29 and September 24, 2005. Since several levees between the city of New Orleans and Lake Pontchartrain were breached, 80% of the cities were flooded. The two hurricanes triggered hazardous material release from on shore and offshore oil and gas installations, with a volume of around 30.2 million liters	Gave opportunity for governments and local people to think how to mitigate the impacts of the future natech disasters
2011	Japan	Earthquake induced tsunami	The Tohoku or Great East Japan Earthquake directly affected many hazardous material releases. It also triggered a tsunami which led to even more damage and destruction among chemical facilities, including the Fukushima nuclear plant. These Natech accidents clearly demonstrated the risk of cascading effects	The necessity of multi-hazard approach was recognized

Europe, and the US. After 2011, the focus has gradually shifted from Natech case studies to Natech measures, such as risk assessment, management, guidelines, databases, and hazard mapping based on European cases, and Natech-specific regulations for risk assessment have recently become effective in some countries. However, raising awareness and including Natech scenarios in these assessments remain insufficient (Hajj et al. 2015). In addition, learning from previous Natech accidents is critical for risk management, and the development of a database is encouraged. "eNatech Database (https://enatech.jrc.ec.europa.eu/)" is one of the examples of Natech accident databases. However, owing to data scarcity and high uncertainty in identifying such hazards, detecting Natech accidents in databases has become a serious challenge for risk managers (Luo et al. 2020).

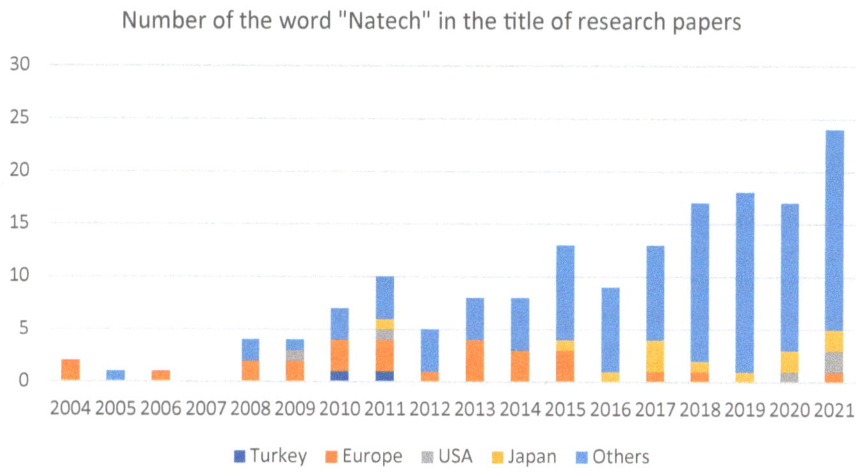

Fig. 8.1 Number of English research papers with the word "Natech" in the title based on a Google Scholar search (*Source* Author)

The category of "others" in Fig. 8.1 includes case studies from multiple countries and Natech topic without "Natech" word in title but in the text. The number of research of "others" has also increased steadily, particularly after the 2010s. The number of non-English papers on Natech accidents has increased, especially since the 2011 Tohoku earthquake-induced tsunami. There are many Japanese research papers on Natech accidents caused by the 2011 Tohoku earthquake-induced tsunami; however, they are not shown in Fig. 8.1.

8.3 Regional Strategies of Natech

The timing of starting research of Natech, its research progress and strategy are different in each country or region. The timing of starting research depends on when the industrialization started in a country. Different strategies have been developed to respond to different natural hazards, and various measures have been developed for each region. The progress and strategies in Europe, the US, and Asia are reviewed, focusing on these differences.

8.3.1 Europe

In Fig. 8.1, Europe has the highest number of English research papers with the word "Natech" in the title, compared to other regions. This is because many researchers in Europe are able to read and write research papers in English and there are many

English journals in Europe. In addition, most European countries are adjacent to one another and share international rivers and the primary natural hazard in Europe is floods (EEA 2010, EM-DAT 2023). Due to the existence of many international rivers, various flood research and countermeasures have been taken in Europe in collaboration with neighboring countries. Especially after the Industrial Revolution began in Britain in the 1830s–1840s, the risk of industrial accidents caused by floods increased, and Natech research has been also taken in collaboration with neighboring countries in Europe. Under such circumstances, a catastrophic industrial accident occurred in a small chemical manufacturing plant in Seveso, Italy in 1976. Thousands of persons were potentially exposed to the highly toxic dioxin, 2,3,7,8-tetrachlorodibenzo-p-dioxin (Bertazzi 1991). Though it is not a Natech accident, it caused tremendous damage such as the long-term effects on health and environment because of the release of the highly toxic dioxin into the air and soil contamination (Bertazzi 1991), and it prompted the adoption of legislation on the prevention and control of catastrophic industrial accidents in Europe (EC 2022). In addition, the Seveso disaster gave rise to numerous scientific studies and standardized industrial safety regulations, such as the European Union's Seveso Directive, which controlled the risks of catastrophic industrial accidents. Due to the inclusion of Natech risks in the Seveso Directive, awareness of this risk has been rising steadily in Europe, especially after Natech accidents during the 2002 summer floods (Table 8.1) (Girgin et al. 2017). In addition, Seveso-III (Directive 2012/18/EU) includes the risk of abnormal weather and asks factories to identify and assess environmental risks, such as floods and earthquakes, in their regular safety reports.

8.3.2 The USA

The main natural hazards that cause technical disasters in Europe and the US, are different. In Europe, most Natechs are caused by flooding. In the US, the majority of Natech accidents are caused by various meteorological phenomena, such as hurricanes, heavy rains, and storms (Luo et al. 2020). Therefore, the US needs to respond to all types of Natech accidents. The natural hazards in each state are different, and each state prepare for and respond to the local natural hazard-induced disasters. If the scale of a disaster is large and each state cannot respond to it, federal organization, such as the Federal Emergency Management Agency (FEMA), responds. In 1979, FEMA was established to respond to all hazards, including natural hazard-induced and man-made disasters, and any kind and scale of disasters (Fukami and Hisamoto 2011). It is a permanent administrative organization for responding to disasters by integrating the following six organizations:

- Federal Insurance Administration
- National Fire Prevention and Control Administration
- National Weather Service Community Preparedness Program
- Federal Preparedness Agency of the General Services Administration

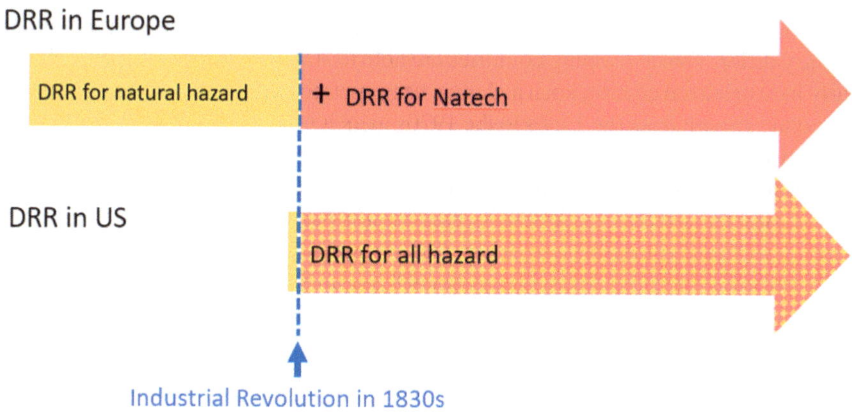

Fig. 8.2 Different disaster risk reduction (DRR) approaches in Europe and the US (*Source* Author)

- Federal Disaster Assistance Administration activities from Housing and Urban Development
- Defense Department's Defense Civil Preparedness Agency.

This approach is different from the European approach that mainly focuses on floods; it rather focuses on responding to all accidents. When they were established, the major disasters to be responded to were still natural disasters. However, the crisis management system changed each time after a critical disaster occurred. For instance, the system shifted to stressing on anti-terrorism after the September 11th terrorist attacks of 1993. Hurricane Katrina also affected the system. In addition, FEMA was directed to lead the operations of COVID-19 by the White House, and it was the first national pandemic response that the FEMA has led since its establishment in 1979 (FEMA 2021).

The disaster-response system developed differently from that in Europe. Europe has a long history before the Industrial Revolution, and the disaster risk reduction system for natural disasters had been developed before the Industrial Revolution. On the other hand, the US became independent in 1776. Only about 60 years after its independence, the Industrial Revolution began in Britain and spread to the US. As a result, the US needed to develop many aspects simultaneously, such as legislation, industrialization, natural hazard-induced disaster risk reduction, and Natech accident risk reduction. Figure 8.2 shows the different process of disaster risk reduction approach of Europe and the US.

8.3.3 *Asia*

While Europe and the US are industrialized countries, many Asian countries are still developing. Depending on the timing of the industrialization in a country, the disaster risk reduction approach and the Natech risk management of each Asian country differ. The developing countries are able to start the all-hazard approach. An early industrialized country such as Japan, has experienced Natech accidents after the economic growth from the 1960s to the 1970s, which was the turning point from light industry to heavy industry. Japan experiences many natural hazard-induced disasters, such as floods, landslides, tsunamis, volcanic eruptions, and earthquakes. Specialists have conducted research separately for each natural hazard-induced disaster, and they have been effectively applied to prevent and mitigate each disaster. As a result, cross-academic research activities involving Natech research in existing research activities have not been highly appreciated. However, Japan has already experienced several catastrophic Natech accidents, such as Tohoku and the Great East Japan Earthquake in 2011. The necessity of a multi-hazard approach to disaster risk reduction has been recognized.

In addition, while Europe and US are continents, many Asian countries, such as Indonesia, Malaysia, the Philippines, and Japan, are islands although some countries, such as Nepal, India, and Bangladesh, are on the continent and share major rivers. Most Asian countries experience various natural disasters, such as floods, landslides, earthquakes, volcanic eruptions, and heavy storms (cyclones and monsoons) while the major common natural disasters in Asia are hydro-meteorological disasters, such as floods caused by heavy rain (EM-DAT 2023). In addition, there are countries with few natural disasters, such as Singapore. Thus, Asian countries are categorized whether it is industrialized or not, and whether it is islands or continents (Table 8.2).

Table 8.2 Natech risk reduction or response approach of Asian countries (*Source* Author)

	Disaster risk reduction (DRR) approach	Geographical feature	Country
Industrialized countries	DRR for natural hazards had already developed before the industrialization, and DRR for Natech and/or other hazards is added after the industrialization	Continent	
		Island	Japan
Industrializing countries	DRR for all hazards have been developing with industrialization	Continent	Bangladesh, Nepal, Myanmar
		Island	Indonesia, the Philippines, Sri Lanka

8.4 Comparison of Different Types and Future Perspectives of DRR

The research progress and strategy of Natech accidents is different in each region or country. European countries have implemented common safety management measures and policies against Natech accidents. In contrast, each Asian country takes separate measures, and research progress varies (Table 8.3).

Table 8.3 Comparison of the Disaster Risk Reduction in each region and country

	Natural hazard	Industrialized/ Industrializing	Continent/ Island	Disaster Risk Reduction (DRR) for Natech and all hazards in progress	
Europe Italy German France, etc.	Flood etc.	Industrialized	Continent and a few islands	DRR for natural hazards had already developed before the Industrial Revolution/ industrialization, and DRR for Natech and/ or other hazards is added after the industrialization	Natural + Natech + other hazards
US	Flood Hurricane etc.	Industrialized	Continent	Both DRR for all hazards and industrialization have been developed and promoted together	All hazards
Asia Japan	Flood Landslide Earthquake etc.	Industrialized	Island	DRR for natural hazards had already developed before the industrialization, and DRR for Natech and/ or other hazards is added after the industrialization	Natural + Natech + other hazards
Asia Bangladesh Nepal Myanmar, etc.	Variety	Industrializing	Continent	DRR for natural hazards has been developing, and industrialization has been promoting	All hazard approach with neighboring countries is expected
Asia Indonesia Philippines Sri Lanka, etc.	Variety	Industrializing	Island	DRR for natural hazards has been developing, and industrialization has been promoting	All hazard approach is expected

Europe and Asia have suffered from natural hazard-induced disasters long before industrialization. The major natural hazard-induced disasters in Europe are rain-induced floods, which are the primary causes of Natech accidents. Such hydro-meteorological hazards are predictable and forecastable and can be prepared for prevention and mitigation. Therefore, flood prevention and mitigation measures, such as flood hazard maps, warnings, and evacuations have been developed for a long time.

Although the most frequent natural hazard-induced disasters in Europe and Asia are floods, there are other natural hazard-induced disasters in Asia, such as landslides, earthquakes, and volcanic eruptions. The other major difference between Europe and Asia is whether the countries are on the same continent. Although most European countries are adjacent to one another, many Asian countries are islands. These locational and geological differences cause various natural hazards. Earthquakes and volcanic eruptions can directly cause industrial disasters. The timing of occurrence of Natech accidents caused by earthquakes is shorter than the occurrence of Natech accidents caused by floods. However, their frequency is much lower than that of hydro-meteorological disasters. In addition, because the prediction of the occurrence of earthquakes and volcanic eruptions is difficult compared to hydro-meteorological disasters, it is difficult to issue warnings for evacuation. Thus, since the research for Natech accidents caused by earthquake and volcanic eruption is less, the research progress is less compared to Natech accidents caused by floods.

The differences between continents and islands also affect the timing of industrialization. This occurred almost simultaneously in European countries. As many European countries share international rivers, Natech accidents induced by floods must be prepared for and responded to in cooperation with neighboring countries. European countries are united in many aspects, such as their economic and developmental degrees. Consequently, it is natural for them to simultaneously deal with Natech accidents.

The timing of industrialization in Asian countries is quite different, as is the degree of development in Asia. As some Asian countries have already developed measures against natural hazard-induced disasters before industrialization, some industrialized Asian countries with various natural hazards have already developed different institutions, measures, and organizations for each natural hazard. This is effective for preparing for and responding to disasters induced by natural hazards; however, it is not effective for Natech accidents or complex/compounding/cascading disasters.

On the contrary, the US could develop an organization for disaster risk reduction for complex/compounding/cascading disasters from the beginning, as the timing gap between independence and industrialization is not large. Thus, this all-hazard approach is possible for US to deal with Natech accidents. This may be one of the reasons why the number of Natech research papers has not increased in the US, as shown in Fig. 8.1. Another reason the number of Natech research papers did not increase in the US is population, land size, and land use. The available living area in the US is larger than that in European and Asian countries. In addition, the timing of industrialization and land development overlapped in the US. As a result, the US could develop a better urban design by separating living areas and factories with toxic substances. If living areas and factories with toxic substances are located far away, the frequency of Natech accidents may decrease. Land size and land use made a significant difference in disaster risk reduction between the US and other small European and Asian countries. Although Europe and Asia focus on preparedness and response, the US focuses on responding.

Since the available living area is limited in Europe and Asia, living areas and factories with toxic substances are nearby. The accidental release of hazardous chemicals has been a long-term concern for owners and operators of industrial facilities, workers, and residents living near industrial facilities (Steinberg et al. 2008). Companies in all countries are asked to prepare and respond to all hazards, and workers in these facilities and neighboring residents also need to prepare and respond to all hazards, including Natech accidents. As a result, owners, operators, and workers of industrial facilities are expected to adopt an all-hazard approach. Because research on Natech accidents is part of the all-hazard approach, cooperation among them and researchers is expected.

8.5 Conclusion

Natech accidents cause huge economic losses as well as human health and environmental problems. Research on natural disasters is essential for risk reduction. Disasters can result from multiple hazards and complex combinations of natural and human-made factors. Therefore, all hazard approaches are expected, especially for the owners and workers of factories with toxic substances and neighboring people, as they are expected to prepare and respond to all disasters. Consequently, multi-sectoral cooperation and national, regional, and multilateral responses are required. Thus, research on Natech accidents is an effective part of the all-hazards approach.

References

Bertazzi PA (1991) Long-term effects of chemical disasters. Lessons and results from Seveso. Sci Total Environ 106(1–2):5–20

Cruz AM, Krausmann E (2009) Hazardous-materials releases from offshore oil and gas facilities and emergency response following Hurricanes Katrina and Rita. J Loss Prevent Process Indus 22(1):59–65

Cruz AM, Suarez-Paba MC (2019) Advances in Natech research: an overview. Progr Disaster Sci 1:100013

European Commission (2022). https://ec.europa.eu/environment/seveso/index.htm

European Environment Agency (2010) Mapping the impacts of natural hazards and technological accidents in Europe. EEA Technical report No 13/2010 ISSN 1725–2237

EM-DAT (2023). https://www.emdat.be/

eNatech (2022) Natural hazard-triggered technological accidents database, European Commission Joint Research Centre. http://enatech.jrc.ec.europa.eu/

FEMA (2021) Pandemic response to coronavirus disease 2019 (COVID-19): initial assessment report. https://www.fema.gov/sites/default/files/documents/fema_covid-19-initial-assessment-report_2021.pdf

Fukami M, Hisamoto N (2011) Education, research and development for risk management in the United States of America: EMI and higher education program. Kyoto University Graduate School of Economics, Working Paper 2011, J-84. http://hdl.handle.net/2433/141782

Girgin S (2011) The Natech events during the 17 August 1999 Kocaeli earthquake: aftermath and lessons learned. Nat Hazards Earth Syst Sci 11:1129–1140. www.nat-hazards-earth-syst-sci.net/11/1129/2011/. https://doi.org/10.5194/nhess-11-1129-2011

Girgin S, Necci A, Krausmann E (2017) 9. Natech hazard and risk assessment. In: Words into action guidelines: national disaster risk assessment, united nations office for disaster risk reduction. https://www.undrr.org/publication/natech-hazard-and-risk-assessment

Hajj CEl, Piatyszek E, Tardy A, Laforest V (2015) Development of generic bow-tie diagrams of accidental scenarios triggered by flooding of industrial facilities (Natech). J Loss Prevent Process Indus 36:72–83. ISSN 0950–4230, https://doi.org/10.1016/j.jlp.2015.05.003. https://www.sciencedirect.com/science/article/pii/S0950423015001242

Krausmann E, Cruz AM (2013) Impact of the 11 March 2011, Great East Japan earthquake and tsunami on the chemical industry. Nat Hazards 67:811–828. https://doi.org/10.1007/s11069-013-0607-0

Kreibich H, Thieken AH, Petrow T, Müller M, Merz B (2005) Flood loss reduction of private households due to building precautionary measures—lessons learned from the Elbe flood in August 2002. Nat Hazards Earth Syst Sci 5:117–126. https://doi.org/10.5194/nhess-5-117-2005

Luo X, Cruz AM, Tzioutzios D (2020) Extracting Natech reports from large databases: development of a semi-intelligent Natech identification framework. Int J Disaster Risk Sci 11:735–750. https://doi.org/10.1007/s13753-020-00314-6

OECD (2022) The OECD programme on chemical accidents started work on Natech risk management in 2008 in partnership with the European commission joint research center, the UNEP/OCHA joint environment unit, UNECE and UNEP. https://www.oecd.org/chemicalsafety/chemical-accidents/risks-from-natural-hazards-at-hazardous-installations.htm. Accessed Dec 2022

Picou JS (2009) Katrina as a Natech disaster: toxic contamination and long-term risks for residents of new Orleans. J Appl Soc Sci 3(2):39–55 (17 pages)

Scawthorn C, Eidinger JM, Schiff AJ (2005) Fire following earthquake. ASCE, NFPA, Reston, Virginia. ISBN 9780784407394

Showalter PS, Myers MF (1992) Natural disasters as the cause of technological emergencies: a review of the decade, 1980–1989. Natural Hazards Research and Applications Information Center, Institute of Behavioral Science, University of Colorado

Steinberg LJ, Sengul H, Cruz AM (2008) Natech risk and management: an assessment of the state of the art. Nat Hazards 2008(46):143–152. https://doi.org/10.1007/s11069-007-9205-3-Springer

Svetlanaa D, Radovana D, Ján D (2015) The economic impact of floods and their importance in different regions of the world with emphasis on Europe. Procedia Econ Finance 34:649–655. Elsevier

Toksoz MN, Reilinger RE, Doll CG, Barka AA, Yalcin N (1999) Izmit (Turkey) earthquake of 17August 1999: first report. Seismol Res Lett 70(6)

Chapter 9
Adopting a Transboundary Risk Management Paradigm in a World of Multi-hazard Risks

Sivapuram Venkata Rama Krishna Prabhakar

Abstract The risks, climatic and non-climatic, are increasingly integrating and interacting with each other across geographical and time scales. This is largely happening due to factors and processes outside the purview of the risk management community that they have mostly ignored for most of the risk management history. The recent pandemic COVID-19 has thrown sufficient light on this new paradigm. Though several previous global incidents have tried to bring to attention this fact, not sufficient attention could be brought to the issue. This is high time for all stakeholders to revisit their risk assessment and risk management approaches and look at all risks in an interconnected manner. Sectorally- and geographically-bounded risk assessments are passed. One may still want to conduct a sectorally- and geographically-limited risk assessment but that only gives a part of the risk picture and needs to be compared with the risk assessments that take a holistic picture of risks across sectors, time series, and political boundaries. Here, the message is not just taking a multi-hazard approach but taking an approach that cuts across political boundaries and time scales as risks accumulate over time and geographical scales. Keeping this in view, this paper makes a case for defining and addressing climate fragility in a multi-hazard and transboundary context. While doing so, it presents various short cases to derive messages relevant to climate fragility risk reduction.

Keywords Transboundary risks · Multi-hazard risks · Disaster risk reduction · Climate change adaptation · Risk management · Risk assessment

9.1 Introduction

In a globalized world, disaster and climate change risks seldom remain within political boundaries but often spill over beyond boundaries. These transboundary risks bring an added complexity to the way the multi-hazard risks impact society. Until

S. V. R. K. Prabhakar (✉)
Principal Policy Researcher, Institute for Global Environmental Strategies, Hayama, Japan
e-mail: prabhakar@iges.or.jp

T. Izumi et al. (eds.), *All-Hazards Approach*, Disaster Risk Reduction,
https://doi.org/10.1007/978-981-97-1860-3_9

now, most risk assessments are focused within one's political boundaries and that worked with limited effectiveness. Without considering the transboundary aspects of risks, any multi-hazard risk assessments will result in the underestimation of risks, and lead to the identification of ineffective and insufficient adaptation actions, resulting in failures in risk reduction. Hence, to strengthen our risk reduction, multi-hazard risk reduction approaches must consider transboundary risks at various stages of risk reduction.

The novel coronavirus disease (COVID-19) could be one of the long-lasting and deadliest pandemics of our lifetime. From the perspective of political boundaries, COVID-19 can be treated as a transboundary risk. The impacts of COVID-19 have been so pervasive that they affected all aspects of our lives including socioeconomic, health, personal security and national security. During the course of the pandemic, it interacted with several other natural hazards including floods, droughts, and typhoons that seriously undermined our ability to manage multiple hazards at the same time. With the pandemic already affecting the capacity of institutions on multiple fronts, the occurrence of hydrometeorological and climatic disasters during the pandemic brought an additional challenge to governments, institutions and people. Before the pandemic, the world experienced another complex disaster of the 2008 global food price crisis. Though termed as a food price crisis, it was a multi-hazard risk that surfaced in the form of a food price crisis. The COVID-19 and 2008 global food price crisis are a few examples of transbondary risks at the global level, and one can find similar examples at the regional and sub-regional levels. These experiences teach us several lessons that are valuable to strengthen our capacity to mitigate multi-hazard risks.

From the above discussion, there is a clear need to understand transboundary risks and incorporate that understanding into multi-hazard risk assessments and management planning. With this in view, this chapter reviews the literature pertinent to transboundary multi-hazard experiences and derives some important messages for informing various stages of risk reduction. The chapter explores how multiple hazards interact with each other in terms of how they manifest impacts on societies and countries, the role of risk communication in conveying multi-hazard risks in a transboundary context, the need for improving the risk assessments in terms of their technical aspects, risk communication and their use in risk management and mitigation. Finally, the chapter looks at some of the climate change adaptation and disaster risk reduction solutions that can address multi-hazard risks in a transboundary context including the options for multi-sectoral and regional cooperation.

9.2 Multi-hazard Risks

The term 'multi-hazard risk' refers to risks emanating from multiple hazards and their interactions in space and time dimensions. The ability of multi-hazard risks to manifest at various spatial and time scales makes them difficult to assess and incorporate into our risk mitigation and management plans. Most geographical regions are vulnerable to more than one hazard, and often these hazards take place at the same time or overlap in such a way that puts additional pressure on the disaster risk management (DRM) systems of that region. Since there is no clear definition of what time and geographical time frames determine such a scenario to be categorized as a multi-hazard scenario, it can be considered as a loosely constructed term that primarily indicates that more than one hazard taking place at the same geographical region. However, how farther these hazards take place both in terms of space and time dimensions is also an important consideration since that determines the capacity needed to manage and mitigate such multi-hazard impacts.

Since managing and mitigating multi-hazards can be challenging, governments and institutions want to be aware of the occurrence of such multi-hazards and understand the future risks from those hazards and plan accordingly. Here, the primary emphasis is on the risk assessment itself i.e. understanding answers to questions such as what is the probability of occurrence of multi-hazards, what are the various permutations and combinations of the multi-hazards, what are the projected risks, and what capacities are required to respond, manage and mitigate the risks. Another pertinent, much more nuanced and probably more complicated question is related to the interactions between hazards in spatial and time dimensions that could alter the risk manifestations drastically. Some examples include a province experiencing drought in part of the province and flood in another part of the province at the same time, drought and floods occurring at the same place at the same time, and drought and floods occurring in different seasons at the same location, and drought and floods occurring at different parts of the province in different seasons. In all these circumstances, the nature and degree of impacts on communities, the environment, institutions and the economy could vary widely. Here, the important assumption we made is that the political boundary is the same i.e. the same prefecture which is reasonably a larger area. However, multi-hazard incidences can occur at the micro-scale level as well. A stark example of this could be a landslide happening in the upper reaches during heavy rainfall even with floods in the lower reaches. The risk assessments ought to capture these multiple dimensions of multi-hazard risks.

Though often a clear distinction is not made between multi-hazard and primary and secondary disasters (cascading disasters) in the published literature, it can be inferred that multi-hazards may or may not be related to each other i.e. a hazard may not have led to the occurrence of the other hazard. However, in the case of primary and secondary disasters (or cascading disasters), there is a clear linkage between the first disaster and the second disaster and hence the secondary disaster may not have happened in the absence of the first disaster. Though is often inferred that the secondary disaster has a lesser impact than the primary disaster, it may not always be the case. For example, if COVID-19 is considered a health disaster, it has caused much more economic damage due to its cascading impacts on other sectors than due to the direct impact on the health of the people. Hence, one may consider COVID-19 as resulting in a much more serious economic disaster than a health disaster even though it is not classified as an economic disaster. The Emergency Events Database (EM-DAT) of The Centre for Research on the Epidemiology of Disasters (CRED) reports the secondary disasters and hence the data can show a part of the picture

Table 9.1 Percentage of secondary disasters associated with primary disasters globally during 1900–2022 (*Source* EM-DAT)

Primary disasters	Secondary disasters	% of primary disasters associated with secondary disasters
Droughts	Famine, food shortage, water shortage, crop failure, wildfire, heatwaves, and epidemics	36.5
Storms	Floods, landslides, avalanches, transport accidents, dam burst, lightening,	29
Floods	Storms, landslides, dam burst, lightening, transport accidents, industrial accidents, crop failure, epidemics, and famine	23.6
Earthquakes	Tsunami, tidal waves, slides, avalanches, fires, floods, and volcanic activity	21.5
Wild fires	Heatwaves, pollution	9.3
Landslides	Floods, transport accidents, industrial accidents,	9.1

Source Author based on EM-DAT, 2022

of the multi-hazard scenario of a region. For example, food shortages, famines, and disease outbreaks during droughts and floods are reported as major disasters. As of writing this chapter, EM-DAT reported 25,762 disasters of various kinds across the world spanning the duration of 1900–2022. Out of these disasters, 3,588 disasters, constituting 14% of all disasters, are reported to be associated with a secondary disaster worldwide. The most associated disasters are floods and storms, droughts and food shortages and famines, droughts and heatwaves, earthquakes and tsunamis, and floods and landslides. Most drought and storm events have an associated secondary disaster followed by floods and earthquakes (Table 9.1). The information on associating disasters will help disaster risk managers to be prepared. Since this information is location-specific, having such information for the local conditions is necessary for taking adequate preparedness measures.

9.3 Understanding Transboundary Risks

Another important geographical dimension is the transboundary or political boundaries. When hazards are separated by political boundaries, the nature of impacts experienced from these hazards across such political boundaries could differ from those experienced within the political boundaries. In a transboundary scenario, the nature of impacts depends on how far the points of source and destinations are separated and how well the political boundaries filter the risks. Just like the previous example, the time dimension also plays an important role in a transboundary scenario.

Transboundary risks refer to the risks originating from across the political boundaries due to the hazards that happen outside the reference political boundaries.

In an interconnected world, the impacts of hazards occurring at a place seldom remain within the boundaries but are bound to spill over across the boundaries through various channels that include natural channels and those that the human society has established. Natural channels include natural resource circulations such as transboundary rivers, biodiversity and ecosystem services etc. Human society has established several channels for their welfare that include distributed manufacturing systems, supply chains, financial markets, trade and investments, human migration etc. Most of these channels transmit risks from one country to another country depending on how well the countries are connected by these channels.

As opposed to transboundary risks, transboundary disasters are disasters that happen in a contiguous area spanning multiple countries or across country boundaries. Hazards such as serious droughts, large typhoons/cyclones and floods can manifest in a large region covering several countries. Several transboundary droughts can be observed in South Asia, South East Asia and Africa.

The transboundary dimension of multi-hazard risks is often ignored for various reasons. First, it is ignored for the reason that most hazard and risk assessments consider political boundaries as an important unit of risk assessment and risks are assessed within a political boundary to assist the administration in taking a decision. These risk assessments often confine the hazard and impact assessments within that political boundary without factoring in the possibility of the hazard happening outside its boundaries. However, political boundaries are often transparent to hazard impacts, with some exceptions of national boundaries, which means that risks can travel across political boundaries with limited hindrance. This makes the regions much more vulnerable to such risks as they do not factor these risks into their risk management plans. The second reason for ignoring transboundary risks is the lack of information sharing across political boundaries. This happens often in the case of national boundaries.

Transboundary disasters can be identified by looking at international disaster databases such as EM-DAT which provides an exhaustive list of major disasters in the world though with little difficulty. The database can help us to identify disasters that span across boundaries as it lists the locations affected by the disaster. For example, the drought of 2009 that affected a major part of India has also affected several parts of Nepal and Bangladesh. Another transboundary disaster during recent years is the forest fires in India and Nepal in February 2016. These events can be easily located in the EM-DAT database as they share the same start and end year and months which indicates the linkage with the failure of the southwest monsoon season or high temperatures (Table 9.2). EM-DAT reports start and end dates for very few disasters. Identifying transboundary risks is not as easy as identifying transboundary disasters as risk transmission pathways are not captured in EM-DAT. Hence, a separate case study of transboundary risks needs to be carried out to better understand them.

Table 9.2 Selected major multi-hazard transboundary events identified in different continents

Transboundary disaster	Period	Multi-hazard and transboundary nature
World food price crisis	2007–2008	A multi-year and multi-country event where natural hazards such as droughts at the time and geographical scales interacted with human interventions leading to the rise of food prices across the world
Southeast Asia drought	2019–2020	A multi-year and multi-country drought that interreacted with COVID-19 and the saline water intrusion as a result of drought
The great global supply chain disruption	2019–2022	Disruption to global industrial supply chains due to COVID-19, typhoons and related flooding
COVID-19 interaction with other climate events	2019–2023	COVID-19 as a transboundary health disaster has interacted with various national and regional climatic events undermining the capacity of societies to effectively manage the risk

9.4 Understanding Transboundary Multi-Hazard Risks

9.4.1 World Food Price Crisis (2007–2008)

The realization that global food prices can be significantly influenced by a combination of seemingly isolated factors, including natural hazard impacts and uncoordinated decisions made by multiple actors dispersed across the world in space and time dimensions, during 2008 and 2011 has signified the new normal that governments and other stakeholders have to deal with. The problem of the 2008 global food price crisis was contributed by factors such as droughts in grain-producing nations and rising oil prices (Jha et al. 2012; Prabhakar 2012). Oil price increases have resulted in a general escalation in the prices of fertilizers, food transportation, and industrial agriculture. The root causes may include the increasing use of biofuels in developed countries and increasing demand for a more varied diet across the expanding middle-class populations of Asia. There were also concerns about the role of increased investments by hedge funds, pension funds and sovereign wealth funds in food commodity derivatives markets leading to major shifts in prices (FAO 2009; United Nations 2011). Policies of some Asian countries have also contributed to the crisis, as in the case of Vietnam, India and the PRC which introduced export bans to control domestic prices (Jha et al. 2012).

The global food price crisis repeated itself in 2011, with a relatively higher magnitude than in 2008, due to a different set of factors contributing to the crisis. These factors include failed wheat harvest in Russia leading to reduced export, reduction in US corn production, and major droughts in preceding years in Russia, Argentine and Australia (United Nations 2011). These food price crises have pushed nearly 44 million people into poverty globally (ibid).

Though several factors have contributed to the 2008 and 2011 food price crises, both events had one common factor i.e. natural hazards such as droughts affecting major food-producing countries thus affecting the global food supply. A game-based simulation conducted to envisage the future impacts of such a food price crisis indicated a significant role played by climate change and related weather vagaries (Fisher and Su 2015). Such an increase in food prices could be catastrophic for many food-insecure countries, especially the rural and urban poor. A survey-based study on household food expenditure indicated that a 20% rise in food prices could push 5.65 million people into poverty in the Philippines and 14.67 million in Pakistan (ADB 2008). A 395% increase in food prices, as a result of climate change, could cause social unrest in South East Asia and Africa (Fisher and Su 2015).

Several short-term and long-term solutions, some off-the-shelf and others needing transformative changes, have been proposed to mitigate and respond to the global food price crisis in the future. These solutions range from the greater need for cooperation and coordination among countries for quick response to developing a real-time monitoring and early warning tool that is based on food demand, supply, agricultural production conditions and climate factors (Table 9.3). Most of these measures need regional cooperation and coordination at a scale and depth that has not been undertaken especially for mitigating the food price crisis.

Several of these proposed measures are either already well-developed or are being strengthened. Though many countries have put in place strong food security policies, these policies are largely not informed of their vulnerability to global factors including those that played a role during the 2008 and 2011 crises. Hence, it is important to revisit the food security policies of countries and strengthen them. Sharing information on risk factors for the food price crisis appears to be a challenging area given that most information is either not present or countries find sharing such information sensitive.

Measures such as regional and global food stocks, price volatility forecasting and early warning mechanisms do not exist yet. Though several countries have national food buffer stocks to stabilize food prices and to disseminate food through public distribution systems as in the case of India; it is unclear how these stocks helped stabilize food prices during 2008 and 2011 since many countries with national food stocks were also affected by the crisis. The Association of Southeast Asian Nations (ASEAN) has put in place regional rice buffer stocks 'ASEAN Plus Three Emergency Rice Reserve' to address food security, poverty and malnutrition problems in the region. However, it was not possible to ascertain the efficacy of this measure as its regularization came into effect only in 2010 though piloting has started since 1979 as a part of the 'Agreement on The ASEAN Food Security Reserve' (ASEAN 2019). Similarly, the ASEAN Plus Three Food Security Information System has been introduced in 2003. However, the system could not project food prices as it is an information system and lacks early-warning capabilities, was based on a limited set of indicators that are limited to the ASEAN region and does not consider global factors (ASEAN 2017).

Table 9.3 Measures to mitigate and respond to future global food price crisis

Category	Measures	References
Mitigation measures	• Strengthening agricultural production systems in vulnerable countries • Expanded support to research and extension services • Revisiting food security policies from the perspective of their vulnerability to global processes • Improving the food self-sufficiency of countries • Uncoupling agriculture performance and weather • Developing food price monitoring and early warning tools • Sharing data on risk factors by countries • Disincentivising and decoupling risky investments in agriculture and food markets • Social safety nets include public food distribution systems, insurance, conditional cash transfer etc • Revisiting vulnerability and risk assessments for integrating relevant global and regional factors • Reducing pressure on agricultural land including from biofuels	Fisher and Su (2015), United Nations (2011), FAO (2009), Prabhakar (2012), Jha et al. (2012)
Response	• Global and regional food stocks • Enhanced cooperation and coordination among countries • Critical thresholds to kick inappropriate response measures • An early warning system based on the food-water-energy nexus	ADB (2008), Fisher and Su (2015), Prabhakar (2012)

9.4.2 The Great Global Supply Chain Disruption

Industrial production has evolved over the years. In the early years of the industrial revolution, industrial production was largely characterized by locally procured raw materials, locally hired human labour and locally procured financial capital with a largely local consumer base. However, the landscape of industrial production has changed over the years due to several factors that include product specialization, resource scarcity, cost minimization and profit maximization, time and geographical advantages, political and social environment advantages etc. The evolution of "outsourcing" has exploded in the way the products are produced at an industrial scale.

The various stages of industrial production have subsequently been differentiated both on geographical and time scales. For example, Boeing procured more than 1.7 billion parts and services from 58 countries for manufacturing aircraft in the United States. Despite advantages, outsourcing and dependency on global supply chains have their disadvantages, including increased direct/indirect exposure to climate-related risks. The recent events such as semiconductor supply chain disruption due to the typhoon damage to Klang port of Malaysia exemplify the risks lurking under the global supply chains (Lim 2021, Leslie 2022). The compounding effects of climatic events and the COVID-19 pandemic further exacerbate the supply chain complexities. While several efforts have been made to understand and address some of the issues associated with global supply chains, one set of issues is either less understood or even less addressed i.e. the globalization of local risks through global and regional interconnected supply chains. This deep dive dwells into how global industrial supply chains are vulnerable to climate risks and how to address them.

The characterization of transboundary climate risks involving global supply chains can be done by characterizing the various components of the supply chain management that can provide entry points for climate risks. Critical components of generic supply chain management include planning, sourcing raw materials, manufacturing, delivery, and returns. Manufacturing is a significant part of the supply chain and hence it deserves a detailed discussion also because the manufacturing processes and entities drive the rest of the supply chain elements to a large extent. Supply chains are not independent of the manufacturing facilities they are serving. In a sense, supply chains reflect the practices and choices made by the manufacturing units it is servicing. And hence, it is of paramount importance to understand the production decisions made by manufacturing units that can contribute to the supply chain risks. Understanding these decisions also provides us with a perspective on entry points of climate risks with implications for supply chain disruptions.

One such decision is the distributed production process which enables companies to bring the production to the source of the need (or market demand) rather than to produce at a single location, store and distribute the products to different markets. Another form of distributed manufacturing is to take the production close to the source of raw material. However, the former kind of distributed manufacturing is more prominent than the latter. With the expansion of consumer demand and markets outside the traditional areas, where the industry has operated for a long period and understood the local risks, more and more manufacturers are eager to install production facilities in emerging markets in developing countries. Entering new markets provided them with the opportunity to expand their market reach to new consumers, and better target the products to new consumers' needs and choices while taking the advantage of local cheap labour and lax environmental regulations.

Part of the decision to expand is also due to the stagnant and saturation state of traditional markets, and production economics that pushed industries to the brink of "expand or perish". The rush to capture emerging markets is also encouraged by the doctrine of "early bird incentive" that made them rush, without fully understanding, to occupy immature emerging markets that are fully ripe with risks. Consequently, this has exposed the production processes to new risks that are not fully understood and addressed. These new risks include local social and political risks, climate risks and climate-related disaster risks. While many developing countries are highly vulnerable to climate change and climatic disasters, they are yet to fully develop institutional systems to address climate risks which is important before inviting such new production facilities. As these new industries are not fully cognizant of local contexts, they are also often isolated from the local society and linkages that are often the forte of locally developed industries. Consequently, the firms have done very little or none in understanding and address the local risks in producer countries. Isolation between the industrial establishments and local societies can be detrimental to their long-term sustainability and the integration must take place through robust corporate social responsibility (CSR) operations.

A closely related phenomenon that has strengthened distributed manufacturing is related to the foreign direct investment (FDI) policies of developing countries. FDI policies have become one of the important go-to economic policies for developing countries to attract foreign investments in manufacturing and technology fields. As a result, the FDI flows have doubled in 2016 compared to 1990, with Asia attracting the most FDI in the manufacturing, infrastructure, energy and transport sectors. FDI provided a win–win situation for the developing countries as it brought new technologies and provided the educated workforce with a gainful employment opportunity. Many of these FDI policies often neither specify the climatic risks nor inform the investors of such risks and precautions to be taken. Often, lax environmental regulations in many of these developing countries have only further compounded the risks. Poor risk communication between the recipient country and the investors is one of the important factors behind the global supply chains becoming exposed to climate change risks. In the context of rapidly growing global investments in Asia, countries that are successful in attracting global investments in manufacturing and service industries are also been seen as highly vulnerable to climatic disasters and related losses (The World Bank 2012).

The third factor that is at least partially contributing to the globalization of local risks is the regional integration processes such as those happening in Southeast Asia which is promoting free trade and mobility in line with the European Union. Enhanced trade and investments have been one of the significant developments of regional integration in Asia during recent decades, and in many parts of the world (ADB 2017). Developing countries in general and those in Asia, in particular, have become producers of goods and services for the rest of the world as a result of which investments in the region have grown significantly (The World Bank 2018). While such regional integration processes are beneficial for countries to tap economic growth potential, they are also leading to the expansion of growth in climate-vulnerable locations including coastal areas that are vulnerable to flooding.

The fourth factor that is an integral part of the supply chain risks, closely related to the third factor discussed above, is the climate change vulnerability of manufacturing units themselves. The geographical location where the manufacturing units are installed and their exposure to climate change risks are the crux of the problem. Due to the absence of climate risk assessments in many developing countries, the exposure of these units to climatic hazards is unknown until an extreme event surprises them. This constitutes the accumulation of hidden vulnerability in many developing countries. Due to a lack of land use regulations, poor standards and by-laws that do not impose strict flood and typhoon resistance construction practices, these manufacturing units are highly vulnerable to climatic hazards such as floods and typhoons.

The physical elements of supply chains can make them vulnerable to disruptions from climatic change and related hazards. A large part of the supply chain exposure includes its infrastructure (roads, ports, trains, telecommunications etc.). Infrastructure development in many developing countries is not well developed and strengthening the infrastructure is still a long-pending developmental gap. Further, many countries, including developed countries, have significant infrastructure that is designed by not keeping in view climate change and related hazards. Ports and roads in particular play a central role in supply chains. However, ports and roads are especially highly vulnerable to climate change due to their physical proximity to coastal areas exposed to typhoons and climate change-related sea-level rise. This makes supply chains highly vulnerable to climate change-related disruptions.

Supply chain governance in a legal sense also deserves attention to understand its vulnerability to climate change risks. Unlike manufacturing establishments that are governed by the law of the land, other parts of the supply chains are not governed by any single entity or institution (the trade aspect that largely is focused on the financial aspects) but are often spread across several laws and regulations depending on the element of the supply chain.

Though supply chain management is largely left to individual manufacturing units, they have a limited ability to fully manage supply chains. Though planning is an integral part of supply chain management and simulation and scenario-based planning helps to consider some of the shocks into consideration, a large part of this planning is often focused on production disruptions due to machinery failure etc. rather than focusing on large-scale catastrophic disruption of low probability and high impact events that affect manufacturing and supply chains. Increasingly, most businesses are focusing on lean manufacturing processes. While lean manufacturing can increase efficiency, incorporation of risk management elements is largely missing and identification of risk elements across the length of the supply chain needs specific attention. However, businesses are increasingly aware of the supply chains and the next section introduces some of the measures currently implemented to strengthen the supply chain resilience to shocks.

9.4.2.1 Thailand Case Study for the Transboundary Risks

A recent example of the globalization of local risks is the Bangkok floods of 2011. This event exemplifies that the climatic disasters in the vulnerable countries will not only impact multinational manufacturing entities in these countries, but they can also have a huge impact on the source countries of these manufacturing entities, their societies and institutions and beyond. Further, it exemplifies that the damage to multinational companies has the highest potential to take the risks across borders.

Climatically, the flood event was caused by strong rainfall events in quick succession due to the strong Southern Oscillation Index (SOI) in 2011 (Gale and Saunders 2013). The year was the wettest in the past 61 years. As a result, the Chao Phraya River experienced a river flow with an estimated return period of 10–20 years.

The total economic damage of the flood on Thailand's economy was estimated to be USD 46.5 billion, which amounts to 1.1% of the national GDP (The World Bank 2012). Insured losses were estimated at USD 12 billion (Swiss Re 2011). Nearly 70% of the total losses were borne by the manufacturing sector due to floods in the concentration of industries in the Ayuthaya and Pathum Thani areas. The Nava Nakorn industrial zone in Pathum Thani alone has 270 manufacturing facilities employing nearly 270,000 workers (Kate 2011). Many of these industries produce electronic parts and components mostly for automotive manufacturers. Estimates indicate that nearly 90% of the total economic losses were accrued to the private sector alone. While the overall physical damages were estimated to be fully recovered, losses such as those related to the loss of production may not be recovered and this event was projected to have continued cascading effects in the following years as well making it a long-term economic recovery process. One of the factors that were attributed to the significant impact on industries was attributed to their concentrated location at a single location. While industrial agglomerations provide advantages such as transportation, infrastructure and efficient coordination, the same may have contributed to the complexities that are less understood and less addressed in establishing such agglomerations (Chongvilaivan 2012; Kimura and Obayashi 2011).

While the event has taken place in Thailand, the impact on Japan was significant. More than 550 Japanese affiliate firms were affected by these floods, and production facilities such as buildings and machinery were severely affected due to floodwaters (Hayakawa et al. 2014). In addition to the direct effects, many Japanese firms engaged in supply chains outside flooded areas were also affected by these floods. Consequently, automobile exports from Japan were severely affected. The year-on-year growth rates of automobile exports from Japan fell. This occurred when Japan was already going through the Tohoku earthquake because of which automobile exports already fell during March-July 2011 with a maximum slump in April (−70%) (Chongvilaivan 2012). In comparison, flood-related automobile exports fell by 20% during December. As a result of floods, the insured losses for Japanese firms alone were estimated to be in the range of 10–15 billion USD, a significant part of which was borne by the Thai insurance companies (Meehan 2012).

As these firms provide supplies to factories in Malaysia, Vietnam and Indonesia and other parts of the world, the production of these factories was also affected due to the shock to the supply chains. Consequently, the event had a significant impact on global industrial production as well with estimates putting the losses to the tune of 2.5% of the annual global industrial production (METI 2012). Such negative shocks on multinational corporations (MNCs) in developing countries are a major issue if the foreign investments in vulnerable and developing parts of the world would have to make any significant difference to the economies of these countries and hence the economic planning and investments by MNCs in developing and vulnerable parts of the world need to be addressed with a multi-pronged approach (Kato and Okubo 2017).

9.4.3 COVID-19 and Climate Change Nexus

COVID-19 has emerged as a global pandemic and has tested countries' preparedness to manage pandemics of its scale.

(a) The pandemic has taken the shape of a transboundary risk since the impacts on one country have spilt over the boundaries of countries affecting other countries.
(b) COVID-19 has also emerged as a systemic risk as the pandemic has deepened its impacts beyond the health of human beings and subsequently affected the livelihoods of people and economies of countries as governments and institutions responded to the pandemic with various measures (Fig. 9.1).
(c) The pandemic has also taken the shape of a compound disaster as it interacted with other natural hazards such as floods and typhoons.

Many of the above-stated negative direct and indirect impacts of COVID-19 were not anticipated by most national disaster management systems as they are not designed to manage well pandemics of this scale. This has called for an urgent improvement in risk management strategies to mitigate and manage future pandemics effectively.

Most developing countries in Asia have revamped their disaster management systems recently as a result of global frameworks such as the Hyogo Framework for

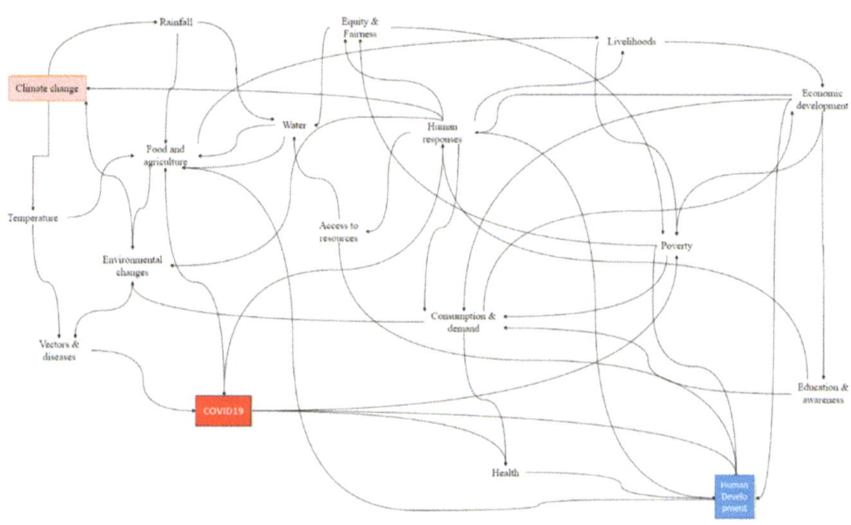

Fig. 9.1 The web of interconnected factors that made COVID-19 a systemic risk (*Source* Authors)

Action and Sendai Framework for Action. However, they have not been designed to manage transboundary risks (TBRs) or pandemics. This leaves the management of transboundary risks such as pandemics to much less known, less maintained and archaic acts, such as the Epidemic Disease Act 1897 of India, which were developed in an outdated context. There is a need to strengthen national and regional risk management strategies to address transboundary, systemic, cascading, and compound risks.

9.4.3.1 Managing Future Pandemic Risks

First, there is a need to improve risk communication at the regional and global levels. Though the evolution of digital communication tools has aided in the effective communication of information among people and institutions, the same has not happened at the macro level. Disease surveillance systems, advanced early warning systems and pandemic preparedness should operate in tandem so that countries can prepare well in advance before a disease takes the shape of a pandemic. Understanding is needed on what causes a particular disease to take the shape of a pandemic, the predisposing factors for the emergence of a pandemic, and the ability to forecast these factors to continuously monitor and manage.

There is also a need for a globally interconnected solution. Ban on the export of medical emergency equipment and medicines by an affected country has put importing countries at risk of a pandemic. Despite the sovereign right of countries to make policy choices, a coordinated response will mitigate the negative impact on others in an integrated world. Formulation of a regional risk management plan will help eliminate such limitations in addressing transboundary risks (Fig. 9.2). Such a plan will bridge the gaps within the national level risk management plan, enhance regional cooperation in risk management and promote complementarity of approaches among countries in a region.

The risk management mechanisms must be significantly coordinated at the federal level, though there has been some progress made at this level already. For instance, whereas many other nations are currently refining their strategies, other countries are effectively coordinating DRM and climate change adaptation. But coordination in other areas of risk reduction is far from adequate. For instance, as evidenced by our experience with COVID-19, the synchronization between the health and DRM systems merits significant consideration. In order for health systems to improve in tandem with the other DRM systems, they must be coordinated with DRM systems at the national to local level. The important phrase here is "extreme event," where the importance of both systems converge. Health emergencies should receive more attention as a result of this collaboration, so that the health systems can be well prepared to address in whatever the form the health emergencies occur i.e. as a consequence of other disasters or in terms of pandemics.

Given the number of pandemics in recent years, national health systems need to be enhanced on par with DRM systems, and they should be interconnected at every level.

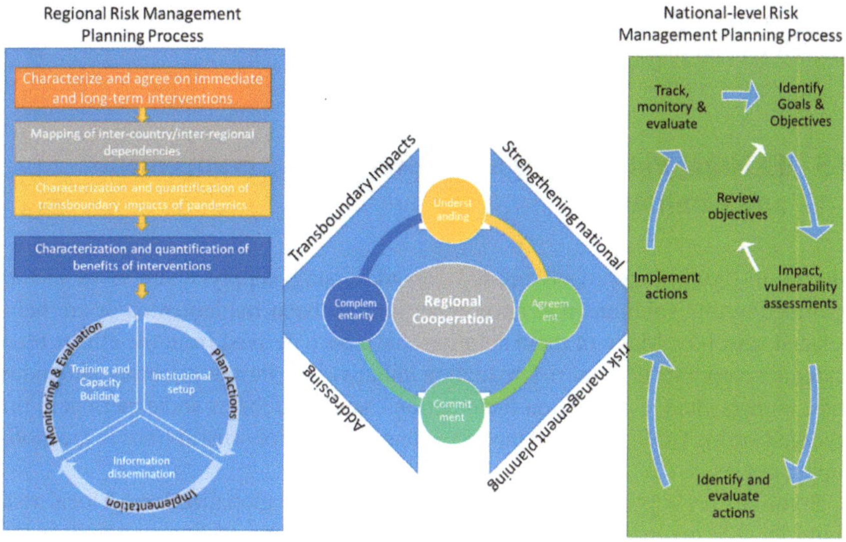

Fig. 9.2 Regional risk management planning framework for addressing the transboundary nature of pandemics (*Source* Author)

Legislation and institutional frameworks for epidemic management that are comparable to national DRM systems are needed for this. It also implies that mandatory emergency drills and pandemic and epidemic simulation games are necessary to be organized by the health departments; governments do not appear to regularly perform emergency exercises or simulation games aimed at handling pandemics. Pandemics should now be a part of national DRM systems' preparedness, emergency exercises and simulation games.

A larger role for the national government is required to support local and regional governments in strengthening their ability to manage pandemics, and various stakeholders in the nation, including local governments, would have to acknowledge the importance of strengthening the capacity to prevent and manage health emergencies, both traditional and non-traditional. This is crucial in nations where national governments have little influence over health-related issues and health is regarded as a policy matter of local governments. When national and local governments can cooperate, putting aside political and jurisdictional issues, many countries can effectively mitigate and manage health emergencies similar to COVID-19 in the future.

Developing the ability of many stakeholders is a crucial component of the plan to create and carry out globally coordinated solutions. More than the governments, the majority of non-governmental organizations (NGOs) have been caught off guard by the COVID-19 pandemic. NGOs typically have a significant impact on disaster management. They can be useful in controlling pandemics in the future if they have the necessary capacity and defined tasks. The public's confidence in government's capacity to contain pandemics must be bolstered by improving health risk governance, which will also enable national and local governments to interact directly with local communities in order to implement policies like self-isolation, self-declaration, and self-quarantine.

9.5 Existing Policy Tools and Governance Arrangements and Limitations

The multi-pronged approach needed for addressing supply chain risks can be informed by the solutions that have either already been implemented or are being debated currently. Many of these solutions have been derived from the efforts made under the corporate disaster risk reduction initiatives by various countries and industries, climate change adaptation interventions, and the debates and experiences that emerged during COVID-19 that have seen one of the serious disruptions to the global supply chains.

Some of the important solutions for mitigating the globalization of local risks could include (a) mandatory risk assessments going beyond financial and market risks to natural hazards, (b) sharing risk information by countries, within the supply chain, and with potential investors, in a transparent manner, and (c) strengthening risk management across the supply chains. These elements are discussed below.

When an investing entity enters into a new market, they are entering not just into a new consumer ecosystem but also into land with new risks posed by climate change to its production and supply chains. Risks such as social risks, and environmental risks including climate change are often not well quantified to be readily used for risk assessments and risk management planning.

Supply chain risk assessments are not new. However, the most current industry practices to mitigate risks largely focus on mechanical failures, labour supply, market risks, resource production and supply, and political risks and largely ignore natural hazards including climatic events. However, this is changing slowly. Many manufacturers are emphasizing the need to understand climate change risks to their supply chains and how these risks interact with other production risks. Studies have also been initiated to understand how the past supply chain risks faced by the major manufacturers and these studies are changing the perspective of the industry significantly. Researchers at Resilinc, the University of Maryland's Supply Chain Management Center and Earth Systems Science Interdisciplinary Center have developed a localized Climate Change Variability/Vulnerability Index that combines long-term temperature and precipitation data and it can help identify climate change risks to individual industrial entities (Boyson et al. 2022). The initiative revealed that nearly 49% of the manufacturing facilities are exposed to climate variability and rapid temperature and precipitation changes. The analysis indicated the possibility of these facilities being affected by floods, storms, heatwaves, water scarcity and fires. Such approaches will help the industry to move away from disaster management that is more focused on event-based planning to a much more long-term and integrated perspective to addressing the risks.

A significant part of the climate change risk to supply chains come from outside the industrial establishments i.e. through damage to roads, ports and telecommunications and many of these are under the purview of governments. Governments also play an important role in incentivizing foreign direct investments and the proliferation of industrial agglomerations and other policy priorities. Climatic events also affect communities from where industries draw their labour force. Hence, supply chain risk management is inherently a multi-stakeholder engagement process. There is a need for the creation of an ecosystem that brings together various stakeholders in a coherent manner to serve the purpose of effective climate risk communication, with contributions from the private sector, governments, civil society organizations and science and technology institutions. Among these stakeholders, the role of governments and the private sector is of paramount importance as direct stakeholders in transboundary investments.

From the point of view of governments, the manufacturing countries need to put in place measures to adapt to the climate risks affecting international investments in specific and the industry in general. National adaptation plans, provincial adaptation plans, and city adaptation plans must focus on industry-specific risk assessments and identify measures where government policies can make a difference. Governments can make it mandatory for all industrial units to conduct their climate change risk assessments, and how it affects the employees and societies that depend on their goods and services. A similar requirement may also be imposed by the industry-originating countries in terms of how the foreign investments made by the domestic MNCs affect national stability and sustainability. Such an overarching requirement could attract resistance in the beginning but the past examples indicate that the industry is capable of adopting stringent measures for its competitiveness and overall sustainability (e.g. environmental standards). Governments would have to facilitate a mechanism where the local climate change risks are well transparently communicated to investors. Likewise, the industry should communicate the supply chain risks that may adversely affect national interests and security. This could be challenging in the beginning as it could disincentivise prospectus industries to invest. However, all things assured, the industry has incentives to work with the governments to address climate change risks and national industrial federations can provide such a gainful interface between governments and the industry.

The investing entities have obligations too. To begin with, they have to be fully aware of the local risks they are entering into by conducting the necessary risk assessments and by broadening the scope of their risk assessments to cover climate change risks. These risks would have to be shared along the supply chain, with the local and national governments and communities whose lives and livelihoods are dependent on the entity and with the shareholders as a part of their investor relations programs. Information shared with shareholders should be beyond the dividends to be paid but also about the climate risk of the investments made by the entity, and measures taken to adapt to the climate change risks.

In terms of planning, business continuity plans (BCPs) have received their fair share of attention in the discourse on disaster risk reduction, and during COVID-19. Several United Nations bodies have developed user guides to educate industries on how to prepare business continuity plans. The focus especially appears to be on small and medium enterprises (SMEs). While BCPs are an essential part of risk management, they are often based on specific short-term disaster events such as fire and floods and are often not based on robust risk assessments that cover long-term risks including climate change. Another important aspect of BCPs is to ensure that these plans are rehearsed and revised regularly and resources are allocated to ensure that the climate change risks are specially addressed.

Constant engagement with input suppliers is an important and emerging area in the industry. Earlier, the companies used to provide design standards and ask the equipment manufacturers to provide components that are subsequently tested and certified by the procuring entity. However, this is changing. More and more manufacturers are engaging with their suppliers on a long-term basis, training them so that the parts manufacturers can test the components even before they arrive at the assembly line. This provides several advantages including avoiding last-minute surprises, reducing delays, dynamically managing the demand, and more importantly improving the trust with parts providers. Boeing through its quality integration strategies, helped suppliers to design and test parts, and by closely monitoring the process, it was able to reduce the delays in input supplies and reduce costs. Some companies such as Mitsubishi are even going one step ahead to advise their parts manufacturers to insure against known risks including climatic disasters.

The redundancy aspect of supply chains has also received much attention in the discourse on supply chain management. Redundancy here refers to identifying and contracting backup manufacturing facilities, multiple-input procurement sources and channels, identifying alternative transportation routes, and constant engagement with the alternative options to understand the lurking risks by making them part of the overall risk assessment and management strategies instead of falling back on them as an afterthought. A closely related aspect is balancing procurement in terms of geographical spread. More and more industries are focusing on procuring locally than depending on distant suppliers, and this is more so after the experiences of COVID-19. Though this can have implications in terms of the availability of suppliers that can meet quality and quantity requirements, corporations can address this by weighing the costs and benefits of building the capacity of local suppliers against the risks involved in procuring from a distance. Application of information technology including blockchain technologies can help maintain such a repository of redundant systems, monitor their health and constantly update the system on their vulnerabilities, help implement instruments such as smart contracts and maintain the overall integrity of the supply chain.

Labour market dynamics and strengthening the social aspects of supply chains is an important aspect of supply chain management. Industries have the luxury of tapping the human resources that are developed by public investments in education and research. This has to change. The industry has the responsibility to engage with the wider public through corporate social responsibility initiatives that address the climate change risks the communities face. This will have a win–win advantage both to the industry as it can buffer the labour forces interruptions while also improving the trust with the society and attracting qualified human resources. This is also the place for the industry to promote equity, diversity and inclusion in its supply chain management. It has been argued that companies have a social responsibility to contribute to and promote the values of equity, diversity and inclusion. Strengthening these values will strengthen the firm's social engagement and social acceptance, build the social capital of the supply chain, and be able to tap innovation and skills from a wider section of society that can make their products and services as appealing to a wider consumer base as possible.

9.6 Conclusion

From the foregone discussion presented in this chapter, it became clear that the world is increasingly becoming a complex place where natural and man-made including climatic events interact with each other at various time and geographical scales. These interactions undermine our ability to effectively understand and manage these events sustainably as our risk governance systems tend to dissect the factors and events effectively delinking the linkages and hence implementing solutions that are not based on the actual interconnected nature of the new and emerging risk landscape that we are increasingly experiencing. To address this, there is a need to change the way we understand and address the risk from the basic perspective itself. A few points can be mentioned here to improve the risk governance from global to local scales: coherence in our understanding of risk, improving the risk assessments and underlying uncertainties, focus on long-term and global risk reduction and effectively breaking down that into shorter and smaller scales without losing the interconnectedness, building the resilience and redundancy in the risk reduction systems, coordinated risk management approaches between and across scales and time dimensions and more importantly making the risk information and the underlying data a fully open to all stakeholders that can revolutionize the risk assessments. The rapidly evolving applications of information technologies including artificial intelligence and artificial neural networks could strengthen our understanding of the interconnected nature of risk across time and geographical scales for identifying solutions at any given point of time and scale.

Acknowledgements The author gratefully acknowledge the resources provided by the following projects that were instrumental in conducting the research and drafting this paper based on the research results: Strategic Research Fund (SRF) of Institute for Global Environmental Strategies, Environment Research and Technology Development Fund (JPMEERF20221C06, 2–2102) of the Environmental Restoration and Conservation Agency of the Ministry of the Environment of Japan (MOEJ) and the Diplomacy and Security Research Project Fund of Ministry of Foreign Affairs (MOFA) of Government of Japan.

References

ADB (2008) Food prices and inflation in developing Asia: is poverty reduction coming to an end? ADB, Manila, The Philippines. Retrieved March 11, 2019, from https://www.adb.org/publications/food-prices-and-inflation-developing-asia-poverty-reduction-coming-end

ADB (2017) Asian Economic Integration Report 2017. ADB, Manila, The Philippines. Retrieved from https://www.adb.org/sites/default/files/publication/375196/aeir-2017.pdf

ASEAN (2017) ASEAN plus three food security information system. Retrieved March 11, 2019, from http://www.aptfsis.org/aboutUs

ASEAN (2019) ASEAN plus three emergency rice reserve. Retrieved March 11, 2019, from https://www.apterr.org/what-is-apterr

Boyson S, Gerst MD, Guntuka L, Linton T, Muraski G, Vakil B, Vakil S (2022) How exposed is your supply chain to climate risks? Harward Bus Rev. Retrieved June 10, 2022, from https://hbr.org/2022/05/how-exposed-is-your-supply-chain-to-climate-risks

Chongvilaivan A (2012) Thailand's 2011 flooding: its impacts on direct exports and global supply chain disruptions. Institute of Southeast Asian Studies, Singapore

FAO (2009) What happened to world food prices and why? The state of agricultural commodity markets 2009. Food and Agriculture Organization, Rome, Italy

Fisher M, Su Y (2015) Food chain reaction: a global food security game. CNA Unlimited, Washington DC, USA. Retrieved March 11, 2019, from http://foodchainreaction.org/wp-content/uploads/2016/02/Technical-Report.pdf

Gale EL, Saunders MA (2013) The 2011 Thailand flood: climate causes and return periods†. Weather 68(9):233–237

Hayakawa K, Matsuura T, Okubo F (2014) Firm-level impacts of natural disasters on production networks: evidence from floods in Thailand. Institute of Developing and Japan External Trade Organization, Tokyo, Japan

Jha S, Roland-Holst D, Sriboonchitta S (2012) Policies for long-term food security in the Greater Mekong Subregion. In: International conference on GMS 2020: balancing economic growth and environmental sustainability. Focusing on food-water-energy nexus. ADB-GMS, Bangkok, Thailand, pp 175–197. Retrieved March 11, 2019, from http://www.gms-eoc.org/uploads/wysiwyg/events/GMS2020-Final-Proceedings/2.1%20Session-Food%20Security.pdf

Kate PT (2011) Thai floods cripple over 300 Japanese firms -JETRO. (REUTERS) Retrieved June 16, 2022, from https://www.reuters.com/article/thailand-floods-supplychain-idUSL3E7LH0GK20111017

Kato H, Okubo T (2017) The impact of a natural disaster on foreign direct investment and vertical linkages. Institute for Economics Studies, Keio University, Tokyo, Japan

Kimura F, Obayashi A (2011) Production networks in East Asia: what we know so far. Asian Development Bank Institute, Hong Kong, China

Leslie J (2022) Climate change will make supply chain woes worse. (Hakai Magazine) Retrieved June 04, 2022, from https://hakaimagazine.com/news/climate-change-will-make-supply-chain-woes-worse/

Lim J (2021) Malaysian floods disrupts semiconductor supply chain; devastates workers. (Techware Asia). Retrieved Junr 16, 2022, from https://techwireasia.com/2021/12/malaysian-floods-devastate-workers-disrupts-semiconductor-supply-chain/

Meehan R (2012) Thailand floods 2011: causes and prospects "om an insurance perspective. Satanford university, Stanford, USA

METI (2012) Floods in Thailand that caused a significant impact on trade environment, etc. of neighboring nations/regions, including Japan. METI, Tokyo, Japan

Prabhakar S (2012) Climate risks to agriculture/food security in the gms countries and early warning systems in the context of the food-water-energy nexus. In: International conference on GMS 2020: balancing economic growth and environmental sustainability. Focusing on food-water-energy nexus. ADB-GMS, Bangkok, Thailand, pp 209–224

Swiss R (2011) Natural Catastrophes and Man-Made Disasters in 2011. Swiss Re, Zurich, Switzerland

The World Bank (2012) Thai flood 2011: overview. Rapid assessment for resilient recovery and reconstruction planning. The World Bank, Bangkok, Thailand

The World Bank (2018) DataBankWorldwide Governance Indicators. The World Bank, Washington DC, USA

United Nations (2011) The global food crisis. United Nations, New York, USA. Retrieved March 11, 2019, from https://www.un.org/esa/socdev/rwss/docs/2011/rwss2011.pdf

Chapter 10
Complex Disaster Risk Management in Afghanistan

Takeshi Komino and Nejabat Khan Safi

Abstract The chapter focuses on the case from Afghanistan where significant part of its population has faced multiple risks particularly recently. On top of continuing high poverty levels, Taliban takeover in the summer of 2021 led to significant reduction in foreign funding, which the situation was exacerbated due to consecutive country-wide drought. The chapter will examine situations and vulnerabilities specific to food security, climate change, and conflict in Afghanistan, key interventions by the humanitarian sector, and proposes future interventions with All Hazard Approach perspective in such context.

Keywords Afghanistan · Disaster risk reduction · Resilience

10.1 Background and Intersection of Multiple Risks in Afghanistan

More than 4 decades of conflict, starting from Soviet invasion in 1979, civil war between political parties in early 1990's, and the Taliban control in Afghanistan has resulted to destructions of institutions, systems, and basic service provision in Afghanistan. As a result, extreme poverty, food insecurity, displacement and forced migration of millions of the people became inevitable. Two decades long US and NATO led operation in Afghanistan also led to the death of tens of thousands of civilians in combat, and wide-spread dissatisfaction towards chronic impoverishment and never-ending conflict situation could be few of the many reasons that led to Taliban re-take-over of the control of the country in August 2021. Afghanistan is facing numerous threats from various terror groups which continue to negatively affect the whole society.

T. Komino (✉)
CWS Japan, Tokyo, Japan
e-mail: t.komino@cwsjapan.org

N. K. Safi
Community World Service Asia, Kabul Afghanistan, Afghanistan

T. Izumi et al. (eds.), *All-Hazards Approach*, Disaster Risk Reduction,
https://doi.org/10.1007/978-981-97-1860-3_10

Renewed international sanctions since 2021 halted vital international assistance that was providing foundational services in many aspects of Afghan people's lives including areas around health and food security. The continued insecurity also encouraged well-educated Afghans to seek other opportunities abroad, which added increased pressure to already-fragile system in the country. As a result, the country's commitment and investments in areas towards SDGs, Sendai Framework for DRR have stagnated.

The economy of Afghanistan is shrinking dramatically by even more than 20%, as it was predicted in UNDP's socio economic outlook (UNDP 2021). The banks are facing serious liquidity shortages, and there are many restrictions on deposit, withdrawal, transfer due to freeze of the country's assets abroad. Afghanistan's GDP was already shrinking in 2020, and it significantly worsened in few years that followed. It led to further deterioration of employment rate in the country, severely reducing already meager coping capacity the general population had towards shocks and stresses, including recurrent disasters. It is increasing vulnerability and reducing resilience which led to food insecurity, poor health condition and access to health services, and malnutrition of all but particularly children and women. Such situation has drove many to adopt negative coping approaches such as joining insurgency groups, drug smuggling, illegal migration, and child labor. Gender based violence has increased, and early and forced marriages for quick cash has increased.

Afghanistan is a landlocked mountainous country with dry climate and with elevation of up to 7,000 m in altitude in central Asian region. The country with its unique landscape also has large dry deserts with areas with little vegetation, and most of the people live in rural areas with their livelihood dependent on subsistence farming. Drought, floods and heatwave are the major climate induced hazards in the country which has its adverse effect on food security, livelihoods, and health of the population, and they influence significantly on surrounding environment including water resources and irrigation, economy, desertification, and wider ecosystem. Absence of clear policies, risk governance mechanism, and poor resources to mitigate the impact of change in climate makes Afghanistan very vulnerable to the climate induced risks.

Afghanistan faces chronic drought. Its intensity and impact increased in recent years due to vulnerabilities such as heavy reliance on rain-fed agriculture of the communities and ecosystems, which left the people highly exposed to the impact of drought. In Afghanistan context, a severe drought generally occurs from low winter rain/snowfall in two successive years, and it has become a recurring phenomenon in recent years. In the last two decades, Afghanistan has experienced more droughts than ever before. Half of Afghanistan's agricultural land depends on spring rainfall, which has become less reliable. According to Food and Agricultural Organization (FAO) in their Afghanistan Drought Risk Management Strategy publication, annual drought in many parts of the country will likely become the norm by 2030 and by 2050 nearly 90% of the country will be affected by drought. The drought in 2018 affected more than two-thirds of Afghanistan (22 out of total 34 provinces) with around 10.5 million people (out of the total 17 million in these 22 provinces) severely affected in their socio-economic well-being and livelihood continuity. As per Integrate Food Security Phase Classification, 13.5 million people were facing 'crisis' category or worse levels of food insecurity in September 2018 and at least 300,000 people were internally displaced due to this drought. UNDP in 2021 gave an estimate that in the worst case scenario, 97% of Afghans could be plunged into poverty by mid-2022.

Flood is another high and widespread climate induced risk in Afghanistan. The World Bank indicates in their Disaster Risk Profile on Afghanistan (The World Bank 2017) that 'floods are the most common type of natural disaster in the country with an average annual financial loss of $ 54 million'. Common type of such floods are flash floods and river floods. Heavy rain and less absorption capacity of earth due to drought, loss of plantation by both human and nature induced causes, increase in surface water, which is compounded by faster melting of snow due to increased temperature. Additionally, inadequate levels of drainage in urban areas also lead to urban flooding as well. Afghanistan's mountainous regions are also exposed to the risk of glacier lake outburst floods (GLOFs), which occur when the natural dam is created which holds meltwater from glacier until it bursts. Recent analysis show that areas with such GLOFs risk is increasing particularly in northeastern regions of the country. These climate risks compound negative impact on already impoverished communities in Afghanistan.

Despite Afghanistan being an agriculture dependent country with 80% of livelihoods directly or indirectly depend on agriculture, historically, the country also depends on food import from neighboring countries such as India, Pakistan, Iran, Kazakhstan, Uzbekistan, Russia, Ukraine, etc. Wheat is the staple crop in Afghanistan, but the country has relied significantly on imports from neighboring countries; according to Ministry of Agriculture and Livestock (MAIL) in its media interview, it revealed that out of annual need of about 6.4 million metric tons, in a normal year Afghanistan import more than 2 million metric tons from different countries of the region (Tolo News 2021). Drought in 2018 reduced wheat yields by more than 60% (The World Bank 2021), and recently, based on Food Security and Agriculture Cluster's reporting, the food basket (wheat flour, rice, cooking oil, pulses, salt) price has increased by 33% between August 2021 and September 2022 (FSAC 2022).

Limited access to agriculture technology, insufficient water system investment, and high population growth rate led to high food insecurity in the country with 11 million people on average experiencing acute food insecurity (Integrated Food Security Phase Classification—IPC Phases 3 and above) even before the recent crisis after August 2021. This number has deteriorated rapidly and by March to May 2022, nearly 20 million people, or close to half the country's estimated population, were classified under high and critical levels of acute food insecurity. Included in this number were 6.6 million people who were classified in Emergency (IPC Phase 4) and 200,000 people in Catastrophe conditions (IPC Phase 5), particularly in Ghor province of the country.

Afghanistan therefore is in one of the worst crisis level in its history, with factors that deteriorate people's vulnerability and coping capacity, while natural and human induced risks increase. It is wide acknowledged fact that it is extremely difficult, nearly impossible, for humans to influence emergence of nature induced hazards, as it is impossible to stop the heavy rainfall or to control the weather events from happening or not happening. Therefore, only ways to reduce disaster risks are to decrease vulnerability, or increase coping capacity. In fact, the vulnerabilities and coping capacity is cross cutting elements for risk reduction from any hazard the country faces, while hazards and exposure to hazards can differ. Unfortunately, Afghanistan is not making enough progress in neither decreasing vulnerability, nor increasing coping capacity in recent years due to prolonged conflict and insufficient level of investment to enhance resilience of the communities as compared to rising level of disaster risks.

10.2 Disaster Risk Management in Afghanistan: Recent Evolution

In order to enhance its disaster governance, the country adopted National Strategy on DRR which was launched in November 2018. This Afghanistan Strategy for Disaster Risk Reduction was developed in consultation with ANDMA's DRR stakeholders from relevant government departments, UN agencies and NGOs during 2017 and 2018 based on Sendai Framework for DRR's key priority areas. The strategy includes subsequent action plans, and was endorsed by Chief Executive Office of Afghanistan's former government.

The strategy aims to guide and lead integration of disaster risk reduction and management in all stages of development process and sectors, with following 5 objectives:

1. Increase integration and cohesion of disaster risk reduction, adaptation to climate change and its mitigation, managing ecosystem, conflicts, fragility and all development aspects that ensure achievement of sustainable development goals and relevant global frameworks and protocols.
2. Long term capacity building including coordination mechanisms at regional and global levels for supporting implementation of Sendai Framework and systematic collaboration for creating flexibility/adoptability against disasters.
3. Strengthening mechanisms, frameworks and capacities at national, provincial and local levels for generalizing, implementation and coordination of DRR strategies and programs which also include risk factors that cause for poverty, unhygienic practices, climate change, vulnerability, establishment of informal/unplanned cities/townships and deterioration of living environment.
4. Systematic inclusion and appropriate use of risk reduction measures in design and implementation of disaster response, preparation and rehabilitations.
5. Establishing tools and equipment for working, and also attracting resources to help implementation of DRR projects and programs.

In order to achieve above objectives, the strategy emphasizes its focus in 5 strategic areas including establishing/creating professional disaster management system, mainstreaming DRR in national development plans and polices, developing institutional and human resources capacities, empowering disaster risks prone communities through improving knowledge and innovations, and expanding DRR program through public and private investment.

Afghanistan National Disaster Management Authority (ANDMA), formerly Department for Disaster Preparedness (DDP) in Afghanistan, was established in 1973 as a coordination office to support a multi-agency National Commission for Disaster Management. The DDP initiated a number of processes and systems, including preparing a Disaster Management Framework, developing the Afghanistan Disaster Management Plan in 2003 and launching a Disaster Management Information System. In 2012, the National Disaster Management Law was enacted and have clearly defined the different types of disasters in the country. The following are key objectives the law.

- Prevent possible natural and human induced disasters and reduce the risk which will caused by natural and unnatural disasters.
- Protect various types of properties, public welfare and safety against natural and human induced disasters.
- Prevent land destruction, conservation areas, vegetative cover and wildlife;
- Setting up, coherent and coordination of disaster management activities at different stages:
- Prepare the people in implementation of disaster management Plan;
- Establish a practical mechanism to combat natural and unnatural disaster based on urgent need of communities.
- Protect the health of disaster affected victim and prevent them from further victimization; and
- Support the women, children, elders, persons with disabilities and those places, which are most vulnerable against disasters.

The law also defines the roles and responsibilities of the National Commission of Disaster Management. Similar structures have been developed at Provincial and district level with the participant of the key line ministries at all level. Law also suggests for financial mechanism for management the disasters.

The ANDMA has 7 directorates at central level which include directorate of finance and administration, directorate of plan and polices, directorate of international relationship, directorate of protection and risk reduction, emergency response, directorate of demining and etc. ANDMA has also established its provincial Directorates in all 34 provinces to regulate activities related to disaster response, preparedness and risk reduction. As such, Afghanistan gave its effort in strengthening its disaster governance through strengthening its institutions.

In addition to above institutional arrangements, there were efforts to address food security measures in Afghanistan as well. Over the past two decades, international community has focused on enhancing agriculture and irrigation in the country, such as the key initiatives illustrated below.

- Food/cash for food assistance, food for education and food fortification: Since last 20 year donors along with UN agencies and NGOs yearly provide food and cash for food assistance to millions of people including internally displaced persons

(IDPs), returnees and disaster affected people in the country. Food were also provided in schools to also (in mean time) to give incentives for further enrollment, wheat flour has been fortified with Soya to increase nutritional intake.

- Rehabilitation of irrigation systems or watershed management activities: More than 300 only international NGOs, UN agencies, Ministries such as Ministry of Rural Rehabilitation and Development (MRRD) with the support of World Bank through National Solidarity Program (NSP) worked for rehabilitation/construction of small to medium irrigation systems which includes diversion dams, water harvesting reservoirs, irrigation canals, traditional underground water system called Karezes,[1] check dams, culverts, aqueducts and other irrigation infrastructures during last 20 years in the country through cash and food for work programs, drought and flood mitigation and response and also under DRR specific interventions.
- Protection of agriculture land through protection walls: during last 20 years NGOs, UN agencies, various Ministries worked for protection of agriculture land from floods and landslides and have protected thousands of hectors of lands throughout the country.
- Construction of dams and major canals: Despite Afghanistan's potential for water resources, only few dams were constructed in history such as Naghlu dam, Darunta dam, Sarubi dam, Kajaki dam, etc. There are however, more potential to utilize water potential in the country with further investments.
- Distribution of improved wheat seeds, fertilizers and agriculture equipment: NGOs, UN agencies, and various Ministries such as Ministry of Agriculture Irrigation and Livestock (MAIL) have proactively provided improved wheat seeds, fertilizers, small agriculture tools such as shovel and wheelbarrow to small landholder formers under different programs. Kitchen Gardening initiative has also been supported through provision of providing vegetable seeds, pesticide and cultivation tool kits, in order to improve household nutrition intake.
- Establishment of orchards: NGOs, UN agencies, and various Ministries have also worked in last 20 years for supporting orchard development of different fruits including apples, apricots, olives, pistachio, almond, etc. The export is still limited, but Afghanistan enjoyed its high quality fruit cultivation.
- Capacity building of farmers: NGOs, UN agencies, and various Ministries also worked with farmers in the country for their capacity building and introduction of best practices around plowing, land preparation, watering to harvesting, storage and improved seeds production of wheat, potato, corn, rice and vegetables.
- Improving animal husbandry through distribution of livestock and animal management trainings: Livestock is a key component in the livelihoods of Afghanistan's largely rural population, particularly for women who play a substantial role in the livestock sector. The initiatives included improving animal husbandry including provision of cattle, sheep, goats and chickens, provision of fodder, as well as

[1] Karezes are constructed on a slope with multiple wells that are connected underground to create a tunnel for groundwater transportation. As water travels through underground, evaporation and loss of water is limited, and it is relatively easy for local communities to maintain the system.

training on fodder preparation, health and hygiene of animals, vaccination and on some of the signs and symptoms of common diseases.
- Establishing and strengthening agriculture cooperatives: NGOs and UN agencies have established agriculture cooperatives and have trained them on different agriculture aspect including collection and distribution of improved seeds, planning of cultivation and type crops, facilitating procurement and supply of fertilizers and pesticide, consultation and connecting farmers with agriculture departments, and marketing skills.
- Construction of cold storages: In some provinces NGOs and UN agencies have also constructed cold storages to store vegetables and fruits for enhanced supply and price controls.
- Improving market skills for agriculture, vegetable and dairy products including processing, sorting and labeling: there were initiatives which worked with farming communities for improving the market skill and exploring market but due to lack of enough resources, equipment, technology, machinery, law local market potential, and limited access to international or overseas markets the skills has not been utilized to reasonable extent.
- Strategies and legal frameworks: NGOs, UN agencies, and various Ministries also worked on developing plans, policies and strategies on food security and relevant aspects of disaster risk reduction.

Some initiatives have sustained, particularly when it was coupled with strengthening local governance and capacity, but it is worthwhile to mention that supply/distribution-oriented assistance style has not led to capacitating the country to manage its shocks and stresses. The efforts to strengthening disaster governance has also been severely affected by conflict, particularly when it comes to implementation of strategies on the ground. The conflict and instability also led to cases of corruption, which sometimes made effective utilization of development funds difficult to achieve the intended results.

United Nations Office for Disaster Risk Reduction (UNDRR) and Asian Disaster Preparedness Center (ADPC) issued a report on Disaster Risk Reduction in Afghanistan (2020) also highlighted the challenges in adequately making progress against Sendai Framework commitments. For example, the challenges illustrated were the conflict, unstable political situation, poor governance, misuse of funds, inadequate level of knowledge to assess needs, vulnerabilities, and for designing interventions and risk reduction initiatives, collecting and utilizing disaster and climate related information. It also highlights the complexity in investing for resilience given so much competing needs, and the critical role of the international community in building resilience.

Afghanistan's effort on DRR was noteworthy, particularly in the last 20 years, but it also had its fragility that it was mainly financed by the international community. Also, effective DRR program shall always embedded in local risk governance, which requires longer term capacity enhancement and localization approach, instead of shorter term emergency response mentality.

10.3 Specific Interventions by Humanitarian Sector and Challenges with All Hazard Approach Perspective

As Afghanistan has continued to be under humanitarian crisis in recent history, there has been considerable focus on humanitarian and emergency response, often associated with provision of food, non-food items, emergency health care, shelter assistance, and education continuity support. These are considered as lifesaving assistance, and one cannot overlook the importance of such interventions.

However, as described in length on Afghanistan's increased vulnerabilities and decreased coping capacity to shocks and stresses, resilience-approach is as much needed as the life-saving humanitarian and emergency response. For example, the authors have initiated disaster risk reduction program with resources and technical expertise from Japan which focused on capacity enhancement in risk assessment, risk communication, risk reduction planning, which included both hard (infrastructure) and soft (skill-based) interventions. From such experience, it is clear that if the communities and local stakeholders understand the science behind why a specific hazard becomes a disaster in their environment, they have the agency to plan and implement mitigation measures. Some of our target communities have spontaneous started to advice returnees where flash flood hazard risk areas are, so that people don't construct their houses in hazard-prone areas, and when and how to evacuate in order to save their lives. There are village leaders who identified significant-sized boulder on the hill that was destabilized due to the earthquake in 2022, and have led the way to demolish the boulder little by little excavation, so that the risk is minimized. In both cases, NGOs the authors are engaged have ensured that science around risk analysis and risk mitigation are communicated to the communities, so that local actions on solutions can take place.

Such success cases are often captured at project-levels, but not at macro, policy-level discussions. The exposure to hazard and associated risks are increasing more than ever due to climate change impact, and countries like Afghanistan need to ensure that resilience-approach is embedded in humanitarian and emergency response programs as well. Aid sector has, and continues to, spend significant resource for life-saving assistance of those communities who are affected by shocks and stresses, including recurrent floods and droughts in the country. As Afghanistan has good agriculture potential with considerable water sources and also most of the families in rural area has 1–2 ha of agriculture land, thus floods and drought risk reduction and adaptation of farming practices in context of DRR and climate change including watershed management, and improving associated institutional capacity and legal framework.

It is important to highlight that health, nutrition and economic status of most families in rural areas of Afghanistan is directly linked and proportionate with food security. Dr. Tetsu Nakamura is a good example of how such approach can transform risks to resilience. Dr. Nakamura initially provided health services in eastern part of Afghanistan, but after years of health services, he reached to the conclusion that the root cause of poor health condition is from malnutrition, which is associated with poor agricultural outputs due to poor irrigation infrastructure. Thus, the medical doctor turned his life-mission into constructing irrigation canals in the area, which led to a transformation in food security in once an arid region in Nangahar province. The communities were empowered, and they have led their own problem solving through their agency. Dr. Nakamura unfortunately was attacked and killed in 2019, but his legacy still remains in many hearts and minds of Afghan people.

In countries like Afghanistan where both chronic and sudden onsets risks are imminent and overwhelming, resilience-approach is the only way to come out from negative spiral of increase in vulnerability due to recurrent disasters. Below are some of the key resilience needs of Afghanistan in coming years, classified into 4 priority of action of the Sendai Framework for DRR:

- **Understanding risks**: the science behind why a hazard becomes a disaster needs to be unpacked and communicated to the communities, so that movement of resilience is generated in Afghanistan. There are reasons why crops fail, landslides happen, flood devastate crop and residential areas. Once the communities understand the science behind what's happening around them, appropriate context specific mitigation measures can be planned and implemented, including monitoring of risks and early safety actions.
- **Strengthening risk governance**: institutional capacity at both local and central levels on risk reduction need to be looked into, as considerable loss of knowledge and technical expertise occurred during the recent transition of the regime. The disaster science need to be reflected to hazard specific guidelines (such as drought guideline), so that risk monitoring and timely risk reduction investments can happen throughout the country. Anticipatory action to mitigate the impact of forthcoming drought is also an important risk governance action area for Afghanistan.
- **Investment into resilience**: local capacity on adaptive farming practices and agricultural inputs need to be enhanced coupled with renewed investment into irrigation system. The infrastructure needs to be monitored and maintained by stakeholders engaged in local governance, such as community development councils. Therefore, capacity enhancement in monitoring and maintenance of key water infrastructure will be critical.
- **Build back better**: life-saving assistance should incorporate resilience factor so that local risk governance is strengthened as a result of humanitarian and emergency response. For example, after each emergency response operation, the affected communities should gain enhanced understanding of why the specific disaster has occurred, and what led the hazard to turn into such a disaster. If all emergency response ensure such risk analysis and communication, the communities can plan how to reduce their risks as part of their recovery plan.

In realizing the above, the principles of localization and partnership is particularly important. Risk identification and planning for mitigation measures should be led by the community representatives and local stakeholders, and outside agencies should focus on providing scientific and technical support in order to realize the aspiration that comes out from local communities. Afghanistan belongs to Afghan people, and they have the agency to determine what issues they need to tackle in order to build their resilience against the specific risks they face.

10.4 Conclusion

The chapter examined the case from Afghanistan where significant part of its population has faced multiple risks, both human and nature induced. Chronic poverty has exacerbated vulnerabilities and reduced coping capacity to shocks and stresses, and on top of such negative trends, climate induced risks including frequent drought and flood have led to significant levels of food insecurity.

In such environment, it is utmost important to apply resilience-approach that genuinely builds or enhances local capacity in assessing risks and mitigating the impact through both locally-led initiatives. The hazard and the exposure will intensify in coming years for Afghanistan without a doubt. Therefore, the question is posed how Afghanistan is going to cope with such a trend, and how it can ensure that it doesn't automatically lead to increase in disaster risks. As described earlier, the only way for this is either reducing vulnerabilities or increasing coping capacities, to be done in the principles of localization and partnership. If such resilience-approach is taken with appropriate level of risk analysis in each community, the residents will be able to plan ahead what specific initiatives are required for safer environment. Ultimately, such understandings and actions will lead to a societal movement which the impact goes far beyond than a hazard-specific project intervention. And only then, All Hazard Approach will be realized in a country like Afghanistan.

References

Food and Agriculture Organization (2020) Afghanistan drought risk management strategy (2019–2030). https://www.fao.org/policy-support/tools-and-publications/resources-details/ru/c/1366257/

Food and Agriculture Organization (2022) Afghanistan: minutes of FSAC monthly meeting (21 September 2022). https://reliefweb.int/report/afghanistan/afghanistan-minutes-fsac-monthly-meeting-21-september-2022

Integrated Food Security Phase Classification (2022) Afghanistan IPC acute food insecurity analysis: March–November 2022 (Issued in May 2022). https://reliefweb.int/report/afghanistan/afghanistan-ipc-acute-food-insecurity-analysis-march-november-2022-issued-may

Tolo News (2021) Govt estimates wheat yield down 20% due to drought: govt. https://tolonews.com/business-173142

United Nations Development Programme (2021) Afghanistan: socio-economic outlook 2021–2022. https://www.undp.org/afghanistan/publications/afghanistan-socio-economic-outlook-2021-2022

United Nations Office for Disaster Risk Reduction (2020) Disaster risk reduction in Afghanistan status report 2020,. https://www.adpc.net/igo/category/ID1656/doc/2021-q74Xpc-ADPC-UNDRR_Afghanistan_Report.pdf

World Bank (2017) Disaster risk profile Afghanistan. https://www.gfdrr.org/sites/default/files/afghanistan_low_FINAL.pdf
World Bank (2021) Climate risk country profile Afghanistan. https://openknowledge.worldbank.org/server/api/core/bitstreams/28ece250-47dc-5181-b937-79980e208853/content

Chapter 11
A Planetary Health Approach to Managing Emerging Risks

Jemilah Mahmood, Saidatul Maisarah Faiesall, and Oliver Lacey-Hall

Abstract The changing nature of disasters through the increased frequency and intensity of natural hazards underscores the need for an all-hazards approach to disaster risk management. This paper proposes for planetary health considerations to be embedded within the all-hazards approach, recognising the inextricable link between human well-being and the health of the planet and its ability to sustain human survival for the years to come. A planetary health approach emphasises the importance of recognising and addressing anthropogenic drivers of environmental destruction that triggers an increase in disaster risks and vulnerabilities. This is uncovered in the paper through presenting and discussing two mutually connected feedback loops between governance of the planet's finite resources and governance of socioeconomic systems at country and global institutional levels. In essence, the paper proposes for planetary health considerations in disaster risk management to both inform and be rooted in shifts in how governance works by generating greater and more focused political will for both mitigation and adaptation measures which must include investing in nature-based solutions, emphasising communications designed for community action, and transforming education systems to contribute towards building community resilience against rising disaster risks.

11.1 Introduction

Humanity has long been afflicted by the burden of navigating uncertainties (UNDP 2022). From conflicts and wars, to plagues, floods, and droughts, building resilience through learning from these disasters has become an underpinning element of human development. The momentous events during recent years have however tested this resilience; human-led activities have increasingly come to shape how the planet functions in what is now known as the Epoch of the Anthropocene, a time in which humanity is the most significant force shaping the functioning of planet Earth and changing the very nature of disasters (UNDP 2022; IPCC 2022; Whitmee et al.

J. Mahmood · S. M. Faiesall (✉) · O. Lacey-Hall
Sunway University, Petaling Jaya, Malaysia
e-mail: maisarahf@sunway.edu.my

T. Izumi et al. (eds.), *All-Hazards Approach*, Disaster Risk Reduction,
https://doi.org/10.1007/978-981-97-1860-3_11

2015). While natural hazards have historically posed the biggest threat to humanity, for the first time in planetary history, humans are driving the existential threats to our existence by destabilising our natural ecosystem and driving social, political and technological transformations that reinforce polarisation (UNDP 2022). In the words of the United Nations Secretary-General, Antonio Guterres,

> Humanity is waging war on nature. This is suicidal. Nature always strikes back – and it is already doing so with growing force and fury. Biodiversity is collapsing. One million species are at risk of extinction. Ecosystems are disappearing before our eyes. Human activities are at the root of our descent toward chaos. But that means human action can help to solve it. (United Nations 2020).

An all-hazards approach towards building resilience against disasters in the age of Anthropocene counters the assumption that disasters are inevitable and independent of human impacts on the natural ecosystem (IPCC 2022). Interactions between the natural environment and humanity are rapidly evolving and becoming increasingly intertwined; disasters are no longer singular "independent" events but instead are compounding to produce major cascading consequences to societies (Levine et al. 2007). The COVID-19 pandemic gives humanity a clear glimpse into what our future may look like. Human-led destruction of the environment through widespread deforestation, wildlife trafficking, unsustainable and unethical food production and other unsustainable practices is bringing humans closer to stressed wild animals that shed viruses, causing a zoonotic spill-over to humans, the consequent rise of infectious diseases, and the potential of such events expanding into epidemics and pandemics (Ford et al. 2022; Özdemir 2020).

Concurrently, we are living against the backdrop of rapid, uneven and chaotic social transformations and technological change that now form a new "uncertainty complex" (UNDP 2022). Humanity faces multi-layered threats of food insecurity, droughts and heatwaves, biodiversity loss, mass population displacement and more frequent and intense infectious disease outbreaks and pandemics, on top of wealth and social inequalities, debt-driven financial crises, intentionally destabilized politics, civil unrest, and a dysfunctional global health system (UNDRR 2019). These impacts all stem from unjust, unequal and imbalanced interactions amongst ourselves and between us and the planet, feeding into the "multiple breadbasket failure" of sudden and gradual tipping points (UNDRR 2019).

The reality that we must now confront is that the health of the planet and the health of humanity are interdependent (Özdemir 2020). This chapter therefore aims to highlight that an all-hazards approach to building resilience and managing disasters must reconcile the anthropogenic nature of future disasters, the interconnectedness of people and the planet, and an emphasis to rethink disaster risk governance to account for rising ecological threats and underlying planetary health drivers.

11.2 Planetary Health: A New Approach for the Anthropocene Epoch

Planetary health is defined by the 2015 Rockefeller Foundation-Lancet Commission as "the health of human civilisation and the state of the natural systems on which it depends" (Whitmee et al. 2015). To advance humanity's well-being, the declining state of planet Earth can no longer be ignored. Planetary health expands the scope of public health, an earlier concept which pertains to the improvement of health of people and communities. The concept builds on 'sustainable development', defined by the 1987 Brundtland Commission as "meeting the needs of the present without compromising the ability of future generations to meet their own needs" (United Nations 1987). Planetary health, in this context, recognises that human development without respect for the planet's natural boundaries produces trade-offs that only hamper sustainable development, leading to greater planetary destabilisation, more disasters and crises and consequent challenges to overall human survival.

Since the economic boom of the 1950s, fuelled by post-war reconstruction, humanity has experienced a "great acceleration"; a period of exponential economic growth, driven in no small part by a massive increase in human consumption of fossil fuels, water, fertilisers, and equally significant growth in production of material goods such as plastic and paper (Steffen et al. 2015). The planetary health approach argues that while humankind has made significant progress in a number of areas as a result of the industrial, agricultural and technological revolutions including, for example, reducing child mortality and increasing life expectancy, these gains and their significant positive developmental impacts have been achieved through unsustainable exploitation that is sacrificing the health of the planet (Planetary Health Alliance 2023). These impacts are now manifesting in our arrival at global tipping points that will manifest in systemic collapse of biodiversity, acceleration in global pollution, and depletion of natural resources, with resulting damage to our lived environment and simultaneously, human existence (UNDRR 2019). The Planetary Health Alliance (PHA) links the interconnected and global challenges that we face today, with solutions that must transcend sectors, disciplines, and silos. The Alliance is simultaneously calling for reform of current values, mindsets, and behavioural norms in order to protect and repair the health of human civilisation and the state of the natural systems on which it depends on (Planetary Health Alliance 2023).

11.3 Planetary Health and Disasters

While the literature on planetary health within the scope of disaster governance remains fairly new, two critical arguments have found a strong foothold in the space. Firstly, planetary health is rooted in the recognition of the existence of the Anthropocene epoch. Impacts from this epochal shift are manifesting in the violation of six out of the nine planetary boundaries that govern our ecosystem: (1) climate change,

(2) freshwater change, (3) pervasive land-system use, (4) reconfiguration of biogeochemical cycles, (5) disruptions in the biosphere integrity, and (6) the rise of novel entities such as plastic pollution (Fig. 11.1) (Rockström et al. 2009; Myers 2017).

Violating one or more of these boundaries places humanity at increased risk of exposure to disasters and crises, while the violation of all nine means global collapse from irreversible environmental changes which will negate humanity's ability to sustain life on Earth as we know it (Myers 2017; Steffen et al. 2015). For instance, biodiversity damage, chemical pollution and more novel entities such as microplastic ubiquity are feeding into and augmenting the scale and impact of climate change (Symeonides 2021). Humanity's transgression of the freshwater change boundary places the world at risk of a global water crisis where two-thirds of the global population may face water shortages by 2050, on top of accelerating species extinction and exacerbating mass displacement, conflicts, and health threats (IEP 2022; Myers

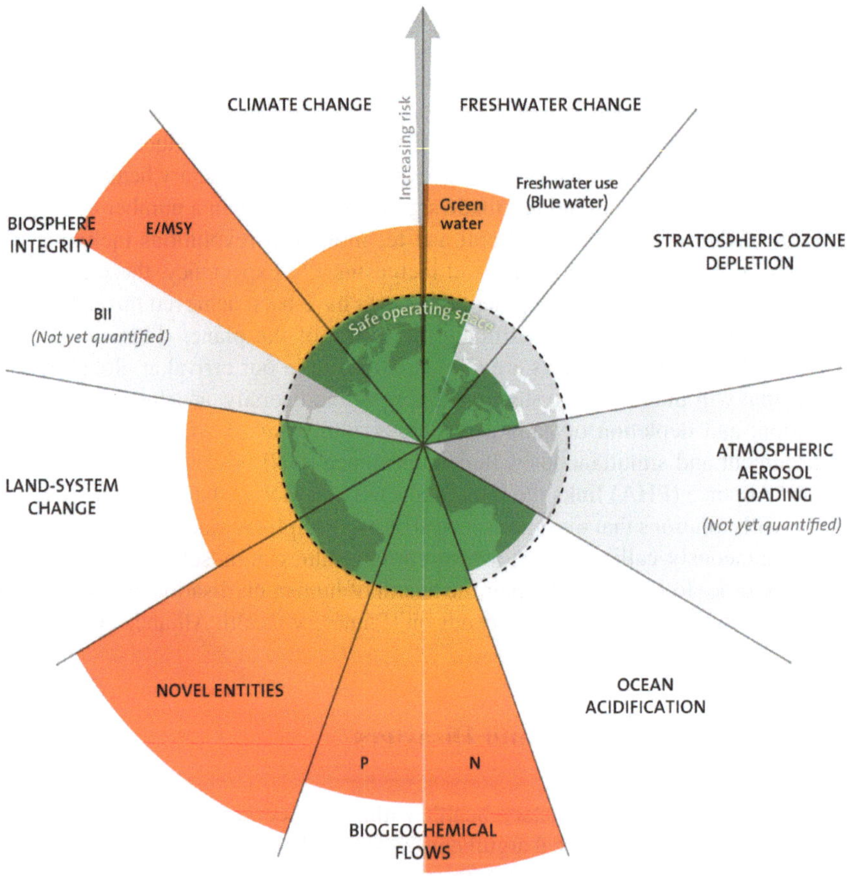

Fig. 11.1 Most recent assessment (2022) of the planetary boundaries

2017). Unchecked land use changes have also increased soil exposure and degradation, exposing more areas and people to disaster risks (Su et al. 2021). This, alongside profligate chemical use is generating biophysical changes that in turn change the temperature, precipitation and moisture levels of soil, thus altering its risk of exposure to more infectious diseases (Altizer et al. 2013). These examples directly contest the misguided belief of disasters being solely "natural". Maintaining healthy natural ecosystems is therefore vital to reducing communities' vulnerabilities to disasters, and in providing a safe operating space for human existence (UNDRR 2022; IPCC 2021).

Secondly, literature on planetary health largely points to it being an approach to systems thinking, rooted in governance (Myers 2017; Pongsiri et al. 2019; UNDRR 2022). Planetary health as a contemporary transdisciplinary practice faces the challenges of imagining a different way of engaging with the planet and translation of related science to support this shift into policy (Özdemir 2020; Myers 2017; Pongsiri et al. 2019). The current structure of our economic systems, that is based on imbalanced norms and power relations, continues to reward and incentivise environmentally degrading industries and flawed metrics such as the notion of limitless economic growth as the main indicator of human progress (Friel 2023; Haines 2016). Operating within the confines of these structures and institutions prevents humanity from making the required shift towards planetary health equity (Friel 2023), and simultaneously diverts attention and resources away from effective crisis mitigation, adaptation, and response efforts, which pose a serious threat to our existence (Haines 2016). The irony is that an all-hazards approach that operates within a flawed system still leaves humanity at the tipping point of systemic failure. In just the first two years of the pandemic between 2020 and 2022, for example, humanity lost five years of previous gains in health and life expectancy, education, and standards of living (UNDP 2022). If we continue business-as-usual in our reactive silos without the urgent proactive transformation of systems and values towards equitable and ethical distribution of finite resources, then there can be little doubt of the increased risk of setting progress back and erasing possibilities to rectify our mistakes. The disaster landscape now is such that despite humanity's best efforts towards achieving the targets of the Sendai Framework 2015–2030, high-frequency, low-impact events continue to set us back significantly (UNDRR 2022).

Offset against this somewhat negative dynamic is a growing recognition of the need for a planetary health discourse and related shift towards a multi-hazards approach. The Association of Southeast Asian Nations (ASEAN) Disaster Resilience Outlook (ASEAN Secretariat 2021) recognises the need for a "necessary lens" and a unifying approach to effectively implement the Sendai Framework, achieve the SDGs by 2030 and ensure a resilient ASEAN by 2035. Still, there is a clear need for more research on the nexus of planetary health and disaster risk reduction, and for future studies to underline planetary health as the basis for disaster risk governance.

11.4 Disaster Governance in the Age of the Anthropocene

The United Nations Disaster Risk Reduction's (UNDRR) Global Assessment Report 2022 recognises an urgent need to rewire systems, leaning on stronger inter-sectoral collaboration and emphasising humanity's value systems and mindsets that should contribute towards risk assessment and resilience-building. Systems must move beyond prevailing reactive disaster responses that focus more on physical infrastructure recovery solutions and population relocation (UNDRR and WMO 2022; UNDRR 2022). With the impacts of the climate crisis becoming increasingly visible, 'governance' in this context includes the recognition that there are two primary feedback loops. The first is between poor governance of the planet's finite resources and increasing social inequalities, political instabilities, and economic burdens. The second involves the inept governance of socioeconomic systems driven by misguided definitions of human development which is now leading to nature biting back.

11.5 Governance of Natural Resources

The COVID-19 pandemic is a clear example of the first loop. Unrelenting encroachment and destruction of the natural environment has led to increasing prevalence and geographic spread of more climate-sensitive and vector-borne diseases such as dengue and malaria (Fletcher et al. 2022; Kulkarni et al. 2022). Mining activity in southern Venezuela, for example, accounted for a 27% rise in malaria incidence (Fletcher et al. 2022). Literature points that malaria transmission in parts of Africa, Latin America, and Southeast Asia are expected to increase due to climate change. Without urgent adaptation strategies globally, other regions such as the United Kingdom are also expected to be climatically suitable for malaria transmission by 2030 (Medlock and Leach 2015; Kulkarni et al. 2022).

It can be argued that we have brought the pandemic upon ourselves—costing our economies trillions of dollars, testing and overwhelming our health systems, eroding trust in political leadership, further widening social inequalities, and leading to millions of deaths worldwide (Australian Institute for Disaster Resilience 2021; Reuters 2022). On average, "natural" disasters such as floods and storms have cost governments worldwide approximately $200 billion per year for the last decade, and this is expected to increase as the frequency of these disasters rise (IEP 2022). Ignoring the immutable link between the behaviour of people and our organising, manufacturing and consumption institutions and the health of the planet leads to heightened risk of disasters, climate events, disrupted food systems and future pandemics (Pongsiri et al. 2019; Nanthini and Gong 2022). Discourse and action to improve disaster risk management governance must include a systems approach that recognises the emerging uncertainties and realities that occur when natural hazards meet man-made disaster risks (Clark 2015; Pongsiri et al. 2019).

11.6 Governance of Socioeconomic Systems

For the second loop, misguided governance of socio-political systems triggers new, more frequent, and more intense natural hazards. To illustrate how one single event can contribute to this feedback loop, the Sichuan earthquake in China is a useful example. In 2008, the government's pursuit of rapid development and economic growth through the rapid construction of physical infrastructure included sacrificing building standards and quality of building materials in pursuit of high-speed construction (Liu and Huang 2014). As a result, these buildings were far from being resilient to the earthquake of 12 May which caused 87,476 casualties, economic damage of USD$ 85 billion, and later cascaded with debris blocking rivers, creating flash floods, and displacing entire villages (Liu and Huang 2014).

Humanity's desire for rapid industrialisation and urbanisation using the flawed metric that is unrestrained GDP growth at the expense of the planet and human health are most evidently illustrated in South and Southeast Asia. Jakarta's overuse of groundwater has led to the city sinking between 1.8 and 10.7 cm annually between 2019 and 2020, with 25% of the city forecasted to be underwater by 2050 (Garschagen et al. 2018). Excessive urbanisation and improper waste management are now also contributing to massive fluvial floods that harm the city's mangrove ecosystem (The World Bank 2011; IEP 2022). Similarly, rapid construction to accommodate the growing population in India has led to significant compromises in the health of its population (Kumar et al. 2015; Rajak and Chattopadhyay 2019). Short-term and long-term exposure to ambient air pollution are responsible for health problems ranging from asthma, bronchitis, pneumonia, deficit lung function, as well as a general overall higher rates of hospital admission and visits (Rajak and Chattopadhyay 2019).

When human development neglects environmental sustainability considerations and is guided only by the blind pursuit of economic growth, the impacts can be deadly, especially in situations where they are multiplied through collision with natural hazards (Douglass and Miller 2018). For example, widening social and economic inequalities in recent years compounded by the emerging seriousness of the climate crisis and the sudden arrival of the COVID-19 pandemic more than doubled acute food insecurity to 345 million people affected (World Food Programme 2023). 2022 was the year of unprecedented hunger with the resultant potential to unleash more conflicts, death and destruction (World Food Programme 2023). Generally, countries with the greatest social imbalances and inequalities will face the most destructive effects of planetary degradation (UNDRR 2022). The cascading nature of these events, both geographically and temporally highlights how humans no longer live independently and confirms, beyond doubt that disasters will continue to compound. Thus, the urgency with which we transform how we manage and mitigate disasters, must be undertaken in lock step with the now inevitable transition in how we live, work, produce and consume.

Quite simply, at the end of this thread is our health, well-being, and survival. Recent events, including those illustrated above, have significantly increased people's vulnerabilities to the spread of diseases and in the worst cases, led to a significant number of deaths as evident during the pandemic. The World Health Organisation (WHO) estimates that one out of four deaths globally are a result of exposure to environmental threats (World Health Organization 2016). Between 2030 and 2050, it is expected that 250,000 people will die from climate-related health impacts including malnutrition, malaria, diarrhoea and heat stress (World Health Organization 2021). Asia is at the heart of this difficult intersection between rising rapid urbanisation and the impacts of the climate crisis. On top of dealing with the COVID-19 pandemic, the region is being further tested with climate and geophysical risks. An unprecedented heatwave combined with an exceptionally heavy monsoon season in Pakistan in 2022 led to melting glaciers that resulted in massive floods and the spread of water-borne diseases, affecting 33 million people and accounting for 1,739 deaths (ESCAP 2023). Annual heat-related deaths in Southeast Asia, including Malaysia, could increase 295% by 2030 and 691% by 2050 (Honda et al. 2014). The planet's health is clearly at risk with devastating and far-reaching consequences to human health; so much so that these climate-related disasters are expected to undo the last fifty years of human development (World Health Organization 2016).

11.7 Planetary Health as an Approach to Managing Disaster Risks

11.7.1 Urbanisation and Country-Level Challenges

At the country-level, the COVID-19 pandemic has given governments around the world the rare opportunity to re-evaluate their policies and entrenched system behaviours and to begin to incorporate planetary health in their policy scoping (Jones et al. 2021) and has mobilised more discourses on planetary health (Hinchcliffe et al. 2021). As urbanisation magnifies risks to compounding disasters, regions and countries undergoing the highest levels of urbanisation, like those in Asia and the Pacific, face the greatest potential for severe climate-related impacts including direct and indirect health impacts (UNDRR 2022). More than half of all South Asians have been affected in some way or another by climate-related disasters such as droughts, floods and earthquakes, breeding a "new climate normal" for the population (The World Bank 2022). In the Philippines, regular earthquakes and frequent typhoons have led people to expect environmental disasters as part of their lived existence, adding a layer of climate anxiety to their daily lives and diminishing trust in the Government's ability to manage and in the individual's agency to generate mitigating change (Douglass and Miller 2018). Southeast Asia and South Asia are also two of the most vulnerable regions to climate change and sea-level rise (UNDRR 2022). When these climate-related impacts compound with wide social inequalities, the results are akin to the widening inequalities, strains on public resources, and an overall increase in individual and household distress—as observed during the pandemic (Claes et al. 2021).

Interestingly and unsurprisingly, countries that have first-hand experiences of climate change related disasters like the Philippines and Indonesia can see the gravity and urgency of the situation (ISEAS-Yusof Ishak Institute 2021). This raises the question of whether individuals and governments alike are, de facto, adopting a 'wait-and-see' approach to managing risks related to climate crisis related disaster events, and, if so, how intense would the impact need to be to spur both into pre-emptive action. Meanwhile, the prevailing attitude globally is of underestimation and invincibility (UNDRR 2022). Planetary health however recognises that adopting measures that lead to the Great Transition—a sustainable and balanced way of coexisting with the planet—cannot happen without a radical change in our behaviours, our values and our mindsets that inform our economies and socio-political systems, and a cross-cutting collaborative model to address the challenges of today. Progress towards SDG1 (poverty reduction), for instance, has come at the expense of higher global material consumption demands and carbon footprint, thereby setting back SDG13 (climate action) and generating more disaster risks (Kroll et al. 2019).

11.8 Global Institutional Implementation

With COVID-19 and the climate crisis, multi-hazard approaches to disaster risk management are moving centre stage. There is increasing recognition that the changing face of disasters is being led by humanity's transgression of the planetary boundaries and resulting anthropogenic consequences. The international humanitarian system is increasingly engaged in discussions and advocacy related to the humanitarian fallout from the climate crisis, from participation in the Conferences of the Parties on climate-related discussions, contributing to the Intergovernmental Panel on Climate Change, to the development of the Inter-Agency Standing Committee's (IASC) Climate and Environment Charter in 2021 and an emerging focus on climate change through establishment of a time-limited IASC Working Group on Climate Change (Inter-Agency Standing Committee 2021). Through consultations with over 200 organisations, the Charter has garnered widespread support from the humanitarian community to begin articulating how best agencies can integrate climate and environmental considerations into their humanitarian and disaster management programming (Inter-Agency Standing Committee 2021; Climate Charter 2023). The Charter, signed by 330 humanitarian organisations around the planet as at January 2023, sends a clear signal of the humanitarian system's changing role and focus, now not only simply supporting people affected by the climate crisis, but focusing also on climate crisis advocacy, climate-sensitive operations and how humanitarian agencies' environmental impacts can be reduced (Climate Charter 2023).

The reality is that humanitarian hotspots are also climate hotspots, and so humanitarian actors have a responsibility to play a part in mitigating the negative impacts of the climate crisis, and addressing the negative impacts of unsustainable humanitarian responses and practices that feed environmental destruction and often undermine people's resilience (IFRC 2019). Directly or indirectly, these impacts range from the large amounts of disposable waste and hazardous substances (plastic sheeting, bottled water, used clothing, non-food items etc.) from humanitarian aid to the carbon footprint generated from air and land transportation, international procurement, water extraction and use and energy consumption, all of which contribute to poor planetary health (Brangeon and Crowley 2020). Between 1994 and 1996, 36 million trees from the Virunga National Park were logged to supply cooking and shelter needs of Rwandan refugees (UNEP and OCHA Environment Unit 2014). More humanitarian actors are recognising their responsibility to include environmental risk and impact assessments in crisis responses to protect the health and well-being of people beyond immediate disaster responses (IFRC 2021; MSF-SIU 2022; MSF 2022). In 2021, the International Federation of the Red Cross (IFRC) articulated the need for a "Green Response" where humanitarian action "saves lives and reduces suffering without risking damage to the environment or the livelihoods, assets, health and survival of affected people" (IFRC 2021). Médecins Sans Frontières (MSF) also now grounds their approach to crisis responses on planetary health, committing to reduce carbon emissions by 50% between 2019 and 2030 (MSF-SIU 2022; MSF 2022).

At the regional level, the focus by ASEAN Humanitarian Assistance Coordinating Centre on Disaster Management (AHA Centre) on climate change adaptation in its Workplan 2025 combined with a focus on sustainable resilience in its upcoming ASEAN Risk Monitor and Disaster Management Review 2023 could serve as a precursor to adoption of a planetary health approach to disaster risk management as the Centre and Association consider ASEAN's forward vision on disaster management and related risk reduction (AHA Centre 2021; AHA Centre 2023). References to cooperation and collaboration with other ASEAN bodies on climate change included in Workplan 2025 are encouraging in this regard.

11.9 New Opportunities for Age-Old Challenges

If there is no change in our current trajectory, evidence suggests that the world is moving inexorably towards a global collapse, led by age-old challenges rooted in, among others, poor and misguided governance, and a lack of transparent and simple communications (UNDP 2022; UNDRR 2022). While disasters are rapidly compounding and cascading, humanity is playing catch-up on addressing the shortcomings of current social, economic and political systems—most notably exposed throughout the course of the pandemic (IPCC 2023). A key elemental change going forward will be building disaster resilience through systemic changes to financial, governance, institutional and policy domains (IPCC 2023; UNDRR 2022; de León et al. 2021). This requirement is not new and is clearly articulated in the various disaster risk reduction frameworks elaborated in 2005 and 2015 (UNDRR 2005, 2015). Planetary health proponents argue that a planetary health approach can support enhanced good governance and social systems and identify ways to ensure better stewardship of natural resources, with the aim of addressing the underlying root drivers of disaster risk exposure that are reinforced by imbalanced power structures and attached to unsustainable economic and political systems, ideologies, and ingrained habits (de León et al. 2017; Steffen et al. 2015). Put simply—planetary health can help connect the currently very disconnected dots.

There are several key challenges to meaningful change in the governance systems that enable effective disaster risk management.

The first is generating adequate political will. Paving a planetary healthy path forward for disaster management means increasing reliance on nature-based solutions and regenerative economics that challenge the deep-rooted dogma of exponential growth (Özdemir et al. 2019). Nature underpins our overall well-being and our economic activities. Half of global Gross Domestic Product (GDP) is dependent on goods and services provided by nature (World Economic Forum 2020). Yet ironically, mankind has built social, economic, and political systems that overvalue narrowly defined GDP growth and undervalue the health and wellbeing of the planet upon which GDP growth ultimately relies.

Nature-based solutions refer to "working with nature to address societal challenges" (Nature-based Solutions Initiative 2023). For example, a flood exposure reduction project in Zambia studied how improved protection against floods through land-use restriction can contribute to economic growth (Joseph 2021). The study found that the project is able to maintain the country's annual average GDP growth at 3% and in the long run, increase it (UNDRR 2022). As more governments begin looking towards nature-based solutions, China has already moved forward in piloting 16 "sponge cities" which are built around a nature-based flood reduction solution (Fu et al. 2022). Urban areas are designed with an abundance of trees, lakes and parks as a new way of promoting urban development, while improving water management, drainage infrastructure, strengthening the natural ecosystem and keeping floods at bay (Fu et al. 2022).

In a ground-breaking move, Malaysia anchored its national blueprint, the 12th Malaysia Plan (12MP) on sustainability and environmental protection with clear references to planetary health, while also starting the process of developing a Planetary Health Action Roadmap for the country (Economic Planning Unit Malaysia 2021; Inani 2022). Meanwhile, Ipoh, a small city in Perak, Malaysia is also reimagining its development plans based on a regenerative economics model—the Doughnut Economics model by Oxford economist, Kate Raworth (Raworth 2017; Rosli 2022). Ipoh City Council has taken the model and identified its own locally adapted set of social foundations to make sure that every development project for the city respects planetary boundaries (Rosli 2022). These case studies show that it is possible to rethink, reimagine, and redesign our socio-political and economic systems so that they are sustainable and respect the planet's natural boundaries.

Secondly is recognising that any sensible way forward involves effective and collaborative communications that emphasise coherence between understanding what the issue is, and the action taken to address or mitigate the issue. As ASEAN has observed, there is a need to break down silos and move from a 'need-to-know' to a 'need-to-share' culture (ASEAN 2021); one that promotes transparency and engages everyone. Humanitarian actors and the disaster management community can play a potentially significant role here—helping to steer the discourse through capturing real-world crisis response case studies including with a focus on trust built between communities in the heat of these crises. Documenting and communicating these observations and experiences is extremely important.

Finally, in terms of education, there is a need to accelerate the introduction of a planetary health approach, and the immutable linkages between people and planet, from an early age (Guzmán et al. 2021). Education systems that expose young children to the risks emanating from climate change produce individuals who are less vulnerable and more adaptable to natural hazards and the impacts of the climate crisis (von Borries 2020). Approaches to communities need to include emphasis on how to navigate through uncertain realities and anxieties, stressing that the future can be positive if transformations are made, and we reconsider our collective and individual approaches to environmental protection.

11.10 Conclusion

Humanity is at a critical crossroads where courage is needed to face up to the fact that human behaviour is the main driver of increasing risks to planetary and human health including, very significantly, disaster-related risk. We need to act urgently to transform our collective existence to one that is based on a truly sustainable growth model grounded in a more equitable, ethical, and responsible distribution of finite resources. Failure to do so means confronting the failures and uncomfortable (at best) consequences of existing socio-political and economic systems that are already manifesting in increasing major compounding and cascading disasters.

The current application of an all-hazards approach to resilient disaster management increasingly recognises the interconnected relationship between people and the planet and builds upon existing public health and humanitarian practices. However, it fails to fully take to account the potential (and arguably actual) environmental impacts of disaster responses. Realignment of the all-hazards approach so that it considers this crucial factor would contribute to systemic change in the humanitarian sector. It would help humanitarian practitioners to recognise and include a key underlying determinant of health and to better understand the changing nature of disasters that are anthropogenic, cascading and compounding, and thus improve the quality and accuracy of preparedness planning.

International humanitarian and disaster management agencies and staff should be encouraged to seize the opportunity to play their part in building resilience to disasters and crises that is grounded in planetary health, recognising the root drivers of disaster risk are not natural phenomena but rather how we mitigate their potential risks, prepare for them, react to them and build back better from them. Making use of planetary health knowledge will generate more informed communities with heightened capacity to mitigate disaster risks. Inevitably, disrupting long-standing and deeply embedded systems and ways of doing things is difficult but not impossible. The bottom line is that there is still hope but it relies on our ability to act now.

References

Altizer S, Ostfeld RS, Johnson PTJ, Kutz S, Harvell CD (2013) Climate change and infectious diseases: from evidence to a predictive framework. Science 2(341):514–519

ASEAN (2021) ASEAN disaster resilience outlook: preparing for a future beyond 2025. https://asean.org/book/asean-disaster-resilience-outlook-preparing-for-a-future-beyond-2025/

AHA Centre (2021) AHA centre work plan 2025: activity, monitoring, and learning plan. The ASEAN coordinating centre for humanitarian assistance on disaster management (AHA Centre). https://ahacentre.org/wp-content/uploads/publications/AHA-Centre-Work-Plan-2025.pdf

AHA Centre (2023) Call for submission: ASEAN risk monitor and disaster management review (ARMOR), 4th edn. https://ahacentre.org/wp-content/uploads/2023/03/2023.02.06-Announcement-of-ARMOR-4th-Edition.pdf

Australian Institute for Disaster Resilience (2021) Systemic disaster risk. https://knowledge.aidr.org.au/media/9228/handbook_systemic_disaster_risk_2022-03-17_v11.pdf

Brangeon S, Crowley F (2020) Environmental footprint of humanitarian assistance: scoping review. INSPIRE Consortium: Humanitarian Policy for Action

Climate Charter (2023) Signatures of the climate charter. https://www.climate-charter.org/signatures/

Claes N, Smeding A, Carré A (2021) Mental health inequalities during COVID-19 Outbreak: the role of financial insecurity and attentional control. Psychologica Belgica 61(1):327–340

Clark H (2015) Governance for planetary health and sustainable development. The Lancet 386(10007):e39-41

De León EA, Shriwise A, Tomson G, Morton S, Lemos DS, Menne B, Dooris M (2021) Beyond building back better: imagining a future for human and planetary health. The Lancet 5(11):e827-839

Douglass M, Miller MA (2018) Disaster justice in Asia's urbanising Anthropocene. Environ Planning E: Nat Space 1(3). https://doi.org/10.1177/2514848618797333

Economic Planning Unit Malaysia (2021) Twelfth Malaysia plan. https://rmke12.epu.gov.my/en/documents/twelfth-plan

ESCAP (2023) 2022: a year when disasters compounded and cascaded. https://www.unescap.org/blog/2022-year-when-disasters-compounded-and-cascaded

Fletcher IK, Grillet ME, Moreno JE, Drakeley C, Hernández-Villena J, Jones KE, Lowe R (2022) Synergies between environmental degradation and climate variation on malaria re-emergence in southern Venezuela: a spatiotemporal modelling study. The Lancet Planetary Health 6(9):e739–e748

Ford JD, Zavaleta-Cortijo C, Ainembabazi T, Anza-Ramirez C, Arotoma-Rojas I, Bezerra J, Chicmana-Zapata V, Galappaththi EK, Hangula M, Kazaana C, Lwasa S, Namanya D, Nkwinti N, Nuwagira R, Okware S, Osipova M, Pickering K, Singh C, Berrang-Ford L, Hyams K, Miranda JJ, Naylor A, New M, van Bavel B (2022) Interactions between climate and COVID-19. The Lancet Planetary Health 6(10):e825–e833

Friel S (2023) The planetary health equity hothouse: a research agenda for positive change. The J Clim Change Health 10

Fu G, Zhang C, Hall JW, Butler D (2022) Are sponge cities the solution to China's growing urban flooding problems? Wires Water 10(1):e1613

Garschagen M, Surtiari G, Harb M (2018) Is Jakarta's new flood risk reduction strategy transformational? Sustainability 10(8):2934

Guzmán CAF, Aguirre AA, Astle B, Barros E, Bayles B, Chimbari M, El-Abbadi N, Evert J, Hackett F, Howard C, Jennings J, Krzyzek A, LeClair J, Maric F, Martin O, Osano O, Patz J, Potter T, Redvers N, Trienekens N, Walpole S, Wilson L, Xu C, Zylstra M (2021) A framework to guide planetary health education. The Lancet Planetary Health 5(5):e253–e255

Haines A (2016) Addressing challenges to human health in the Anthropocene epoch: an overview of the findings of the Rockefeller/lancet commission on planetary health. Public Health Rev 37(14)

Hinchliffe S, Manderson L, Moore M (2021) Planetary healthy publics after COVID-19. The Lancet Planetary Health 5(4):e230–e236

Honda Y, Kondo M, McGregor G, Kim H, Guo Y-L, Hijioka Y, Yoshikawa M, Oka K, Takano S, Hales S, Sari Kovats R (2014) Heat-related mortality risk model for climate change impact projection. Environ Health Prevent Med 19:56–63. https://pubmed.ncbi.nlm.nih.gov/23928946/

IEP (2022) Ecological threat report 2022: analysing ecological threats, resilience & peace. Institute for Economics and Peace, Sydney

IFRC (2019) IFRC secretariat environment policy. July 2019. https://www.ifrc.org/sites/default/files/2021-08/20210813_IFRC-Environment-EN.pdf

IFRC (2021) Green response. https://www.ifrc.org/sites/default/files/2021-11/Green-Response-Intro-Oct-2021.pdf

Inani NF (2022) Planetary health: forging commitment and gaining insights for the development of the national planetary health action plan to drive Malaysia's sustainability. Mahathir Science Award Foundation. https://msa-foundation.org/blog/planetary-health-forum/

Inter-Agency Standing Committee (2021) Consultations on the climate and environment charter for humanitarian organizations: summary of feedback and revisions. Consolidated feedback on the climate and environment charter for humanitarian organizations gathered during consultations conducted between December 2020 and March 2021. https://www.climate-charter.org/about/

IPCC (2021) Climate change 2021: the physical science basis. Contribution of working group I to the Sixth assessment report of the intergovernmental panel on climate change [Masson-Delmotte V, Zhai P, Pirani A, Connors SL, Péan C, Berger S, Caud N, Chen Y, Goldfarb L, Gomis MI, Huang M, Leitzell K, Lonnoy E, Matthews JBR, Maycock TK, Waterfield T, Yelekçi O, Yu R, Zhou B (eds)] Cambridge University Press, Cambridge, United Kingdom and New York, NY, USA, p 2391. https://doi.org/10.1017/9781009157896

IPCC (2023) Synthesis report of the IPCC sixth assessment report (AR6). https://www.ipcc.ch/report/ar6/syr/downloads/report/IPCC_AR6_SYR_LongerReport.pdf

ISEAS-Yusof Ishak Institute (2021) Southeast Asia Climate Outlook: 2021 Survey Report. https://www.iseas.edu.sg/wp-content/uploads/2021/09/Southeast-Asia-Climate-Outlook-2021-Survey-Report.pdf

Jones A, Morelli G, Pettigrew S, Neal B (2021) Report: Integrating wellbeing into the business of government: the feasibility of innovative legal and policy measures to achieve sustainable development in Australia. Victorian Health Promotion Foundation by The George Institute for Global Health

Joseph J (2021) Zambia and Tanzania reap the benefits from DRR investments. International Institute for Applied Systems Analysis. https://www.preventionweb.net/news/zambia-and-tanzania-reap-benefits-drr-investments

Kroll C, Warchold A, Pradhan P (2019) Sustainable Development Goals (SDGs): are we successful in turning trade-offs into synergies?. Palgrave Commun 5(140)

Kulkarni MA, Duguay C, Ost K (2022) Charting the evidence for climate change impacts on the global spread of malaria and dengue and adaptive responses: a scoping review of reviews. Globalization and Health 18(1)

Kumar P, Khare M, Harrison RM, Bloss WJ, Lewis AC, Coe H, Morawska L (2015) New directions: air pollution challenges for developing megacities like Delhi. Atmos Environ 122:657–661. https://doi.org/10.1016/j.atmosenv.2015.10.032

Levine JN, Esnard A, Sapat A (2007) Population displacement and housing dilemmas due to catastrophic disasters. J Planning Literature 22(1)

Liu M, Huang MC (2014) Compound disasters and compounding processes: implications for disaster risk management. Input paper prepared for the Global Assessment Report on Disaster Risk Reduction 2015. https://www.undrr.org/publication/compound-disasters-and-compounding-processes-implications-disaster-risk-management

Medlock JM, Leach SA (2015) Effect of climate change on vector-borne disease risk in the UK. Lancet Infectious Dis 15(6):721–730

MSF (2022) MSF commits to reduce carbon emissions to help safeguard the most vulnerable. 29 March. https://www.msf.org/climate-emergency-msf-commits-reduce-carbon-emissions-help-safeguard-most-vulnerable

MSF-SIU (2022) Planetary health through the MSF humanitarian lens—a snapshot. https://msf-siu.org/blog/planetary-health

Myers SS (2017) Planetary health: protecting human health on a rapidly changing planet. Lancet 390:2860–2868

Nanthini S, Gong L (2022) Planetary health: an alternative framework for disaster governance in ASEAN?. Singapore, RSIS

Nature-based Solutions Initiative (2023) What is the nature-based solutions initiative and what are nature-based solutions?. https://www.naturebasedsolutionsinitiative.org/what-are-nature-based-solutions

Özdemir V (2019) Innovating governance for planetary health with three chritically informed frames. J Integrative Biol 23(12)

Planetary Health Alliance (2023) About the PHA. https://planetaryhealthalliance.org/about-the-pha

Pongsiri MJ, Bickersteth S, Colón C, DeFries R, Dhaliwal M, Georgeson L, Haines A, Linou N, Murray V, Naeem S, Small R, Ungvari J (2019) Planetary health: from concept to decisive action. The Lancet Planetary Health 3(10):e402–e404

Rajak R, Chattopadhyay A (2019) Short and long term exposure to ambient air pollution and impact on health in India: a systematic review. Int J Environ Health Res 30(6):593–617

Raworth K (2017) Doughnut economics: seven ways to think like a 21st-century economist. Random House, London

Reuters (2022) IMF sees cost of COVID pandemic rising beyond $12.5 trillion estimate [online] *Reuters*. 21 January. https://www.reuters.com/business/imf-sees-cost-covid-pandemic-rising-beyond-125-trillion-estimate-2022-01-20/

Rockström J, Steffen W, Noone K, Persson Å, Chapin FS, Lambin E, Lenton TM, Scheffer M, Folke C, Schellnhuber H, Nykvist B, De Wit CA, Hughes T, van der Leeuw S, Rodhe H, Sörlin S, Snyder PK, Costanza R, Svedin U, Falkenmark M, Karlberg L, Corell RW, Fabry VJ, Hansen J, Walker B, Liverman D, Richardson K, Crutzen P, Foley J (2009) Planetary boundaries: exploring the safe operating space for humanity. Ecol Soc 14(2):32. http://www.ecologyandsociety.org/vol14/iss2/art32/

Rosli (2022) Ipoh to be doughnut economic city 2027. Ipoh Echo. https://www.ipohecho.com.my/2022/08/24/ipoh-to-be-doughnut-economic-city-2027/

Steffen W, Richardson K, Rockström J, Cornell SE, Fetzer I, Bennett EM, Biggs R, Carpenter SR, de Vries W, de Wit CA, Folke C, Gerten D, Heinke J, Mace GM, Persson LM, Ramanathan V, Reyers B, Sörlin S (2015) Planetary boundaries: guiding human development on a changing planet. Science 347(6223)

Su Q, Chen K, Liao L (2021) The impact of land use change on disaster risk from the perspective of efficiency. Sustainability 13(6). https://doi.org/10.3390/su13063151

Symeonides C, Brunner M, Mulders Y, Toshniwal P, Cantrell M, Mofflin L, Dunlop S (2021) Buy-now-pay-later: hazards to human and planetary health from plastics production, use and waste. J Paediatr Child Health 57(11):1795–1804

The World Bank (2011) Jakarta: urban challenges in a changing climate: mayors' task force on climate change, disaster risk & the urban poor. Jakarta: The World Bank. November 3, 2011. https://www.worldbank.org/en/news/feature/2011/11/03/jakarta-urban-challenges-in-a-changing-climate

The World Bank (2022) Climate and development in South Asia. https://www.worldbank.org/en/region/sar/brief/integrating-climate-and-development-in-south-asia/integrating-climate-and-development-in-south-asia-region

UNDP (2022) Human development report 2021/2022: uncertain times, unsettled lives: shaping our future in a transforming world. United Nations Development Programme, New York

UNDRR (2019) Global assessment report on disaster risk reduction. United Nations Office for Disaster Risk Reduction, Geneva

UNDRR (2022) Global assessment report on disaster risk reduction: our world at risk: transforming governance for a resilient future. United Nations Office for Disaster Risk Reduction, Geneva
UNDRR and WMO (2022) Global status of multi-hazard early warning systems: target G. United Nations Office for Disaster Risk Reduction
UNDRR (2015) Sendai framework for disaster risk reduction 2015–2030. United Nations Office for Disaster Risk Reduction
UNEP and OCHA Environment Unit (2014) Environment and humanitarian action: increasing effectiveness, sustainability and accountability. https://www.unep.org/explore-topics/disasters-conflicts/what-we-do/preparedness-and-response/environmental
United Nations Secretary-General (2020) The State of the Planet. Address at Columbia University. 2 December. https://www.un.org/en/climatechange/un-secretary-general-speaks-state-planet
United Nations (1987) Our common future. Report of the world commission on environment and development. Transmitted to the general assembly as an annex to document A/42/427—development and international co-operation: environment
Von Borries R, Guinto R, Thomson DJ, Abia WA, Lowe R (2020) Planting sustainable seeds in young minds: the need to teach planetary health to children. The Lancet Planetary Health 4(11):501–502. https://doi.org/10.1016/S2542-5196(20)30241-2
Whitmee S, Haines A, Beyrer C, Boltz F, Capon AG, de Souza Dias BF, Ezeh A, Frumkin H, Gong P, Head P, Horton R, Mace GM, Marten R, Myers SS, Nishtar S, Osofsky SA, Pattanayak SK, Pongsiri MJ, Romanelli C, Soucat A, Vega J, Yach D (2015) Safeguarding human health in the Anthropocene epoch: report of The Rockefeller Foundation-Lancet Commission on planetary health. Lancet 386(10007):1973–2028. https://doi.org/10.1016/S0140-6736(15)60901-1
World Economic Forum (2020) Nature Risk Rising: Why the crisis enguilfing nature matters for business and the economy. https://www3.weforum.org/docs/WEF_New_Nature_Economy_Report_2020.pdf
World Food Programme (2023) A global food crisis. https://www.wfp.org/global-hunger-crisis
World Health Organization (2016) Preventing disease through healthy environments: a global assessment of the burden of disease from environmental risks. https://www.who.int/publications/i/item/9789241565196
World Health Organization (2021) Climate change and health. https://www.who.int/news-room/fact-sheets/detail/climate-change-and-health

Chapter 12
A Systemic Risk Approach for Integrated Disaster Response

Arunabh Mitra, Rohit Chaurasia, and Rajib Shaw

Abstract As the environment and the economy undergo rapid, multidimensional change, disruptive events are becoming more frequent, unexpected, and difficult to plan for. They are often generated by systemic risks with complex, unpredictable, ambiguous, and globalized repercussions, such as pandemics, cyberattacks, climate change, food security, and war, among others. These Systemic risks exceed conventional methods of risk management; therefore they provide unresolved policy and governance issues. The COVID-19 pandemic is one such example, and it has heightened alarm as well as interest in the diversity of such future risks and the complexity of the disruption and impact they may produce. Due to the complexity and mutability of systemic risks, we need a paradigm shift in how we promote the deeper integration of catastrophe risk management across many sectors and levels, taking interdependence, co-benefits, and trade-offs into consideration. Using Covid19 as a case study, this chapter provides an overview of systemic risks, identify key inflection points in its evolution and proposed the basic design principles and conceptual framework for enhancing the overall management of systemic risks from a "Whole Systems Approach."

Keywords All hazards · Systemic risks · Integrated disaster management · Covid 19

A. Mitra (✉)
HCL Technologies, Hyderabad, India
e-mail: arunabh.mitra@gmail.com

R. Chaurasia
Eduvanz Financing Pvt Ltd, Mumbai, India
e-mail: rohit.chaurasia@outlook.com

R. Shaw
Keio University, Shonan Fujisawa Campus, Fujisawa, Japan

T. Izumi et al. (eds.), *All-Hazards Approach*, Disaster Risk Reduction,
https://doi.org/10.1007/978-981-97-1860-3_12

12.1 Introduction

As we strive to build an interconnected world with globalized economies, we have unwittingly contributed to the construction of mechanisms that are capable of not only transforming, amplifying the positive effects but also enabling threats to have a cross geographic effect- thus transforming them into global occurrences, capable of sparking additional global crises.

Systemic risks are those that encompass the environment, society, politics, technology, and economy, each of which poses a unique degree of complexity and requires focused analysis. The pervasiveness of global systemic risks contributes to the evolution of the disaster risk fundamentally; it is now systemic and cascading.

Given the above context, it is necessary to comprehend systemic risks holistically and with a focus on disasters. This chapter discusses the nature of Systemic Risks and proposes the fundamental design principles and conceptual framework for Integrated Disaster Resilience (IDR) based on a 'Whole Systems Approach that will allow us to cut disaster-related losses and prevent the introduction of new threats and serve as a foundation for establishing successful frameworks, policies, and strategies for its evaluation, mitigation, and management.

12.2 Systemic Risk: A Primer

A system as described by the Merriam Webster dictionary is "a system is a regularly interacting or interdependent group of items forming a unified whole. Every system is constrained by space and time, shaped by its surroundings, defined by its structure and function". A complex system is one with numerous interconnected components with its properties including nonlinearity, emergence, spontaneous order, adaptability, and feedback loops that come from these interactions (Complex Systems Journal n.d.). Complex systems are an interdisciplinary area that incorporate contributions from several disciplines, including physics study of self-organization and critical events, social sciences research of spontaneous order, mathematics' study of chaos, and biology's study of adaptability, among others. Environments (like cities), social and economic structures (like ecosystems), and the entire universe are examples of complex systems. Due to dependencies, competitions, relationships, or other sorts of interactions between their elements or between a system, complex systems are difficult to understand.

The domain of Systemic Risk has been evolving for over 70 years. Finding its origins in complexity science- Systemic Risk research began in the early 1950s and accelerated fast (Castellani 2018). The multidisciplinary nature of systemic risk combined other scientific disciplines, such as cybernetics (Weiner), biology, ecology (Understanding the Complexity of Economic, Ecological, and Social Systems 2001; The Paradoxy of Observing Systems 1995), and economics (Systemic risk in banking ecosystems 2011). The 2007–08 financial crisis and climate related impacts piqued interest in systemic risk and its application to real world problems leading to heightened awareness of how specific calamities spread across regions and sectors, causing unanticipated losses and perhaps systemic collapse.

Global connectivity has transformed the frequency, impact and economic tenacity of the risks faced by Human civilization. The OECD was among the first to recognize the emergence of Global Systemic Risks in its 2002 report Emerging Risks in the Twenty-First Century (OECD 2002) followed by The World Economic Forum which was also one of the first to warn about the emergence of systemic risks. It stated that main problem in making systems resilient to such risks is managing their expanding complexities and interdependencies and the need to build resilience in each part of the global system, including finance, supply chains, health, energy, the Internet, and the environment (WEF, The Global Risks Report 2014, 2014).

12.2.1 Systemic Risks in Theory

The two primary theoretical viewpoints form which systemic risk is studied are system theory and network theory (Towards a transdisciplinary approach to systemic risk detection 2021).

12.2.1.1 Systems Theory

Systems theory explores the self-organization of complex structures. It underpins the analysis of systemic risk but has limitations. Key questions such as how complex systems adapt and evolve, the influence of initial conditions on their evolution, the impact of component interplay, and the identification of stabilizing feedback loops are central to this theory. However, challenges include defining system boundaries and pinpointing critical interdependencies and feedback loops. Most approaches in systems theory focus on bottom-up changes, potentially overlooking top-down modifications that might lead to overlooked risks (Arunabh Mitra n.d.).

12.2.1.2 Network Theory

Alternatively referred to as network risk, systemic risk emphasizes the significance of interconnections. Network theory maps a system's structure through its elements' connections, influencing activities like information flow between human agents or the spread of diseases. While integral to understanding systemic risk, network theory has its limitations, such as the tendency to overlook the multi-layered nature of complex systems with overlapping networks. This oversight can lead researchers to erroneous conclusions (Arunabh Mitra n.d.).

12.2.2 Characteristics of Systemic Risks

The four major characteristics demonstrated by Systemic Risks are namely, Complexity, Uncertainty, Ambiguity and Transboundary Effects (Renn et al. 2020). Complexity indicates that the cause–effect relationships between many variables are difficult to discover and measure because they are not evident or readily observable. Uncertainty includes statistical variance, measurement mistakes, ignorance, and indeterminacy (van Asselt, MBA 2000), thus diminishing trust in the predicted cause and effect chain(s). Ambiguity refers to the variety of interpretations that may be drawn from comparable observations as a result of variances in the criteria or standards used to understand a situation; finally, transboundary effects are when a risk can have a big impact outside of where it originated. This means that even though systemic risks start in one part of society or the environment, their effects can be felt in all parts of society, like the economy, politics, and civil society. This ability to go beyond national borders makes multilevel governance necessary.

12.2.3 Governance of Systemic Risks

Governance of Systemic risks is complex and still growing, necessitating a paradigm shift in how we promote greater integration of catastrophe risk solutions across several sectors and levels, taking interdependence, co-benefits, and trade-offs into account. Recognizing the limitations of public and private decision makers in terms of risk and uncertainty is one of the most difficult components of designing risk management approaches to decrease natural disaster losses. The provision of tools and incentives that aid individuals in making more informed and effective decisions is a further obstacle. According to Magnus Hagelsteen (Hagelsteen and Becker n.d.), there are five interrelated challenges that occur when comparing actors' performance in a complex dynamic uncertain environment to capacity building concepts. These are (1) conflicting ideals, (2) irrational control, (3) mindset lag, (4) a lack of willingness for change, and (5) power imbalances.

The 2019 Global Assessment Report (GAR 19) (UNDRR 2019) examines the risks embedded in today's world through a systemic lens and requires the global community to acknowledge and embrace a new appreciation for the dynamic nature of systemic hazards, innovative risk governance frameworks for large adaptive systems, and innovative risk-informed decision-making tools. The paper views the Earth as a single system—a system of systems—and shows why linear, segmented, and reductionist frameworks are inadequate for assessing and managing threats in the 21st Century. The proposed Global Risk Assessment Framework (GRAF) employs multidisciplinary research and analysis to better understand the shifting nature of risk due to the interdependence of our societies and the global economy, where shocks frequently cascade and compound in complex ways with wide-ranging social, fiscal, economic, and environmental repercussions. GAR 19 endorses the International Risk

Governance Centre's (IRGC) principles to manage Systemic Risks. By giving recommendations for managing systemic risks in socio-technical and socio-ecological systems, the IRGC approach to Systemic Risk Governance seeks to enable policy analysts and professionals in the business and public sectors. The advice is intended to assist participants in comprehending complex system dynamics and avoiding or initiating a regime shift, taking into account changes in underlying conditions or proximity to a tipping point that may induce a regime shift. The guidelines are divided into seven interconnected processes for investigating, adapting, and transforming the system in which the organization functions and thereby increasing resilience (Editor, International Risk Governance Council n.d.) Lack of integration or mainstreaming of Disaster Risk Reduction (DRR) into the development planning process is the most significant barrier to DRR adoption.

While significant progress has been made in the planning and implementation of Disaster Risk Governance (DRG) at the international, regional, and national levels, primarily in terms of the formation of organisations and networks for DRR, there is a need for a new style of disaster risk governance in light of the increasing complexity and uncertainty associated with disasters (Arunabh Mitra n.d.).

Global Risks Report 2021 (Editor, Global Risks Report 2021 n.d.) addresses significant governance possibilities for increasing the long-term resilience of governments, corporations, and the international community, with a focus on establishing analytical frameworks that take a holistic and systems-based view of risk impacts. The study Governing Complexity and Its Implications for the Sendai Framework for Disaster Risk Reduction presents five strategies for addressing future complexities and uncertainties in disaster risk reduction would be as follows: reduce underlying vulnerability as a root cause of disasters, be inclusive/leave no one behind, manage urban disaster risks, manage climate change adaptation and mitigation, and govern for resilience: toward adaptive and transformative governance.

Although only a handful of governments have gained the competence to detect, manage, and control novel, unexpected, or complex hazards. In many nations, risk management remains mostly sectoral or thematic, with government bodies directing the process mainly within their areas of expertise, thereby restricting its efficacy. Both institutional risk management strategies and the extent to which non-state entities participate in risk assessments vary considerably. The potential of risk sharing remains largely unrealized, and there is a paucity of efficient feedback loops into the policy cycle.

A new paradigm for systemic risk governance is described for the twenty-first century, with an emphasis on "sustainability and resilience" systemic risk governance should promote reshaping over protecting progress from external risks.

12.3 COVID-19 as a Systemic Risk

In the modern globalized world that is increasingly interconnected, COVID-19 was a dramatic reminder of the ever more complex and systemic nature of risks. The direct and indirect impacts of the pandemic have revealed and reinforced inherent vulnerabilities across societal systems, borders, and scales. They have seen the closure of borders, economic disruptions and failures, strained and overwhelmed health systems, and failure of supply chains, all of which are contributing to ripple effects and human and national security issues (UNU-EHS 2022). Labelled a systemic concern since its emergence- the COVID 19 pandemic resulted in cascading developments that took advantage of the growing worldwide linkages. The overburdening of the healthcare infrastructure by COVID-19 brought forth its systemic elements highlighting the requirement of better coordinated, communal, and multi-dimensional governance (Yong-kyunKim 2022). The pandemic went on to demonstrate how the challenges of dealing with systemic risk go beyond typical risk management and governance. The interdependence of critical systems, as well as their underlying vulnerabilities, has made it more important to know about systemic risks and how to deal with them.

12.3.1 COVID-19: Impact on Systems

What emerged as a health problem engulfed the economy in a nonlinear way, growing rapidly to affect every part of our interconnected existence. COVID-19 has had an influence on almost every socioeconomic area, including the economy, transportation, energy, commerce, finance, and natural phenomena such as climate change and natural catastrophes. Even though there was an application of interventions, a cascading effect on livelihoods, the economy, education, political, and social systems was noted, further exacerbating pre-existing vulnerabilities.

12.3.1.1 Economic Impact

The pandemic had a profound impact on poverty and inequality around the world. Global poverty increased for the first time in a generation, and uneven income decreases among the disadvantaged contributed to a significant rise in inequality within and across nations. The pandemic-induced lockdowns in a number of nations have had a cascading effect on the worldwide supply, resulting in a severe reduction in supply and demand across a wide range of commodities and enterprises, so fostering an atmosphere that is predisposed to a global recession. The combination of actions implemented to handle the short-term consequences differed considerably between nations. In addition to direct income support programmes, governments and central banks made unprecedented use of debt relief policies, such as debt moratoriums for individuals and businesses, decreased interest rates, and loosened liquidity conditions, making it easier for commercial banks and nonbank financial institutions, such as microfinance lenders, to refinance themselves and allowing them to continue extending credit to individuals and businesses. Despite the fact that these programmes alleviated the short-term liquidity issues faced by households and businesses, they had the unintended consequence of obscuring the true financial condition of borrowers, thereby creating a new problem: a lack of transparency regarding the true extent of credit risk in the economy. This financial risk might impact the entire economy through routes connecting the financial health of families, businesses, financial institutions, and governments. (Editor, World Development Report 2022-Chapter 1. The economic impacts of the COVID-19 crisis, 2022) This interdependence can lead a high financial risk in one area to spread and destabilise the entire economy. The adoption of a curfew was extremely disruptive for informal business activity, which led to a decline in income. Leading to a generalized rise in poverty across regions where a complete lockdown was imposed, this impacted the most economically vulnerable segments of the population that are primarily dependent upon the informal economy. This impact was not only faced in urban zones but also rural areas experienced economic disruption due to the disruption of the regular need for market goods, which showed signs of system dependency and rural–urban links in trade networks. Perhaps the most tangible example of this is demonstrated through the slowdown of the global transport and trade networks, resulting in decreases in national productivity and subsequent widespread job losses (UNU-EHS 2022). According to survey data, in 2020 temporary unemployment was higher in 70% of all countries for workers who had completed only a primary education (Kugler et al. 2021). The efforts to contain the pandemic entailed imposing quarantine, preparing health facilities, isolating infected individuals, and tracking contacts, all of which leveraged public health resources, personnel, and implementation expenses- contributing to fiscal stress especially in lower-middle-income countries (LMICs) where fiscal constraints are higher. The economies across the world observed rise in public health expenditure, and revenue decline due to the government's failure to increase revenue due to quarantine and curfews. This economic shock extended to shortage of labor because of illness, rise in mortality, and fear-induced behavior. Other than labor shortages, disruption of transportation, closure of workplaces, restricted trade and travel, and closed land border are reasons for the pandemic's economic slowdown (Yunfeng Shang 2021).

On the flipside, one of the greatest positive impacts of COVID-19 has been the catalyzation of digitization- both economy and workspace. Digital payment usage has surged as a result of the pandemic. Over forty percent of adults in low- and middle-income nations (excluding China) have used a card, phone, or the internet to make in-store or online payments for the first time since the epidemic. More than one-third of adults in low- and middle-income countries paid their power payments directly from their bank accounts. After the outbreak, more than 80 million adults in India and more than 100 million adults in China made their first digital merchant payments, respectively.

Two-thirds of adults around the world now make or receive digital payments, with the proportion in developing economies increasing from 35% in 2014 to 57% in 2021. 71% of people in developing economies have a bank, other financial institution, or mobile money account, up from 63% in 2017 (Editor, COVID-19 Drives Global Surge in use of Digital Payments 2022). Similarly, the pandemic has accelerated trends in remote work, e-commerce, and automation. Remote working has cascaded into denting business travel since the extensive use of videoconferencing during the pandemic a new acceptance of virtual meetings and other aspects of work. This adaption is an irreversible change which can have significant knock-on effects on employment in commercial aerospace, airports, hospitality, and food service (Susan Lund 2021).

While the pandemic presents an opportunity to adjust our economic institutions and enhance their resilience and responsiveness designing the new economic system to be more inclusive, sustainable, and environmentally friendly, overarching global economic conditions will continue to moderate COVID-19's effect on education, work, and training. The global economic contraction caused by the pandemic—and directly influenced by countries' responses to the crisis, along with increased levels of public borrowing, economic uncertainty, a large number of failed businesses, and decreased economic confidence both productivity and investment by businesses have slowed as a result- having potentially far-reaching and long-lasting effects on individuals and society. We hope to convey some of the broader societal impacts of COVID-19 in the next section.

12.3.1.2 Social Impact

Early in the pandemic, community connection increased, with Local and hyperlocal charitable and voluntary organisations becoming essential to the response against COVID-19, however there are disparities between towns based on the quality of their community infrastructures. Resulting in disparities: communities with greater community infrastructure and social capital have performed better, establishing that effective community-led initiatives need to be supported by a well-established and adequately funded community infrastructure. The risks connected with unemployment, failing enterprises, reduced consumption, and major changes in the structure of the economy are anticipated to have a significant impact on tax collections and thus government expenditure in the medium to long term. It will become increasingly important to strike a balance between income production and expenditure versus non-economic externalities, considering the societal goals that are to be achieved. The social impact of the pandemic led closures wasn't restricted to adults and working professionals but found itself extending to children and young people, directly impacting their access to education. Since learning varies according to socio-economic profiles of students and schools, individuals from low-income families were more likely to be the most affected by the closure of schools as well as a transition to digital learning. A 2021 BCG report citied that half of the 1.5 billion students affected by school closures worldwide faced economic and technical barriers to online learning, as reported by UNESCO in April 2020—most of them in developing countries (Bansal 2021). The economically backward strata children also suffered by being deprived of school meals and other social welfare measures based on schools. The impact of COVID-19 extended across to the workforce as well, with employment trends highlighted recent McKinsey Report that spoke of greater changes in the mix of jobs within economies. Compared to the pre-COVID-19 estimates recorded by McKinsey, they expect the largest negative impact of the pandemic to fall on less skilled employees in office support roles, including in food service and customer sales and service roles, while roles in warehousing and transportation may increase as a result of the growth in e-commerce and the delivery economy, however those increases are unlikely to offset the disruption of many low-wage jobs (Susan Lund 2021). That being said, globally, the need for skilled workers in the healthcare and STEM associated fields may expand more than it did before the pandemic, due to an increased focus on health as the population ages with a rise in income, as well as the growing need for individuals who can develop, deploy, and manage new technology.

Another overlooked aspect of societal changes associated to the remote first work pattern is how people's mental health has been affected to varied degrees during the pandemic, and especially during the lockdown period when they were expected to deliver on professional commitments while being left vulnerable to substance abuse or loneliness, having to deal with grief of having lost loved ones to the pandemic in isolation and living in fear of the devastating economic consequences of the pandemic. Exacerbated by inability to treat new occurrences of mental illness and addressing preexisting disorders which are both impacted by lockdown measures including restrictions on access to general healthcare and mental health support services.

12.3.1.3 Environmental Impact

The environment is fundamental to human and animal health. COVID-19 is a challenge to world health in the twenty-first century. In the aftermath of the pandemic, lockdowns, quarantines, and border closures have resulted to a reduction in air pollution due to decreased travel and productivity. Countries across the globe imposing lockdown and the WHO declaring emergency with social distancing measures to combat the novel coronavirus pandemic resulted in the complete closure of various industrial activities, contributing up to a 5 °C decrease in temperature compared to pre-lockdown periods, indicating that the industrial sector is likely responsible for the energy footprints that can dramatically increase global temperatures. (Kumar et al. n.d.) Storms, floods, wildfires, and other natural disasters that cause billions of dollars in damage are aggravated by climate change. Significant near-term reductions in greenhouse gas emissions are required to avert climate risk and keep temperatures below the thresholds that would start runaway warming. A low-carbon recovery strategy could not only initiate the substantial emissions reductions required to halt climate change but also determine whether the world meets or misses the emissions goals of the 2015 Paris Agreement, which were established to limit global warming to 1.5–2 °C to prevent irreversible climate change. Achieving this will necessitate rapid, capital-intensive action in every sector of the economy, while the easing or deferral of environmental regulations in the interest of economic growth, depressed oil prices which will make low-carbon technologies less competitive, and stimulus programmes will divert funds that governments would otherwise invest in a zero-carbon transition. Thus, building economic structure is crucial to the cascading effect of economic consequences on changes in environmental pressure.

Given the absence of demand, the Service and Aviation industries, which were among the hardest hit by the epidemic, produced fewer emissions. Significantly beneficial reductions in environmental pressures during COVID 19 led to a 7% fall in energy-related emissions (Kumar n.d.). Long-term changes in environmental pressure will be largely determined by economic forces and regional implications. Due to the impact on the transportation sector, fossil fuel use, which is linked to greenhouse gas (GHG) and air pollution emissions, was significantly reduced, leading experts to suggest that this unprecedented decrease in air pollutant emissions during the pandemic could reduce seasonal ozone concentrations.

There has been an improvement in the health of the world's bodies of water due to travel limitations and decreased economic activity. As a result of lockdowns having a negative impact on global tourism, fewer ships are traversing global waterways for tourist, shipping, fishing, and coastal aquaculture. This reduction in water travel and activity has decreased marine disturbances caused by noise pollution from ships, fishing sonar, and recreational boats, as well as the possibility of ships striking marine species and harming or killing them. During the pandemic, water ecosystems in many parts of the world showed indications of recovery. During the lockdown, the water quality of the Ganga River, India's "National River," improved, particularly in industrial clusters and urban areas (Chakraborty 2021) Increased dissolved oxygen and decreased biological oxygen demand, faecal coliform, total coliform, and nitrate content were indicators of the effect. With better water quality, the self-cleaning properties of the river were improved. Typically, enormous quantities of untreated or partially treated effluent from home and industrial wastes enter the river, resulting in a catastrophic decline in water quality. Several unsuccessful government programmes have cost the government millions of dollars over the past two decades. During the lockdown period, industrial wastewater decreased by around 1,300–1,340 million litres per day. However, air pollutants could also contribute to nutrient contamination in streams. For example, NO_2 can react with other chemicals to produce acid rain, which is particularly detrimental to freshwater ecosystems such as streams, rivers, lakes, and watersheds.

As a result of the temporary closure of onshore enterprises, the demand for power reduced across all sectors, with the greatest impact felt in the manufacturing sector due to the lower need for metals in industries such as construction and automobile manufacturing. Pharmaceuticals was one of the only industries to increase output due to an increase in demand. However, an often-overlooked source of pollution is the improper management of medical waste generated in part by an increase in the use of disposable PPE kits by patients and healthcare professionals, whose indiscriminate disposal has contributed to the spread of infectious diseases and toxic effects. In areas lacking regulated municipal and industrial waste collection and management services, the likelihood of illegal dumping, open burning, and incineration was greater, contributing to poor air quality and human health due to exposure to contaminants. The incorrect disposal of single-use plastics such as syringes, gloves, and trash in water bodies can result in the leaching of inert polymers into the environment, altering the quality of the soil and water, as well as the animals that consume or come into touch with them. Additionally, plastic waste can provide breeding grounds for insects. For zoonotic disease carriers, such as those that carry Zika and dengue, public health and safety become threats.

The global consequences of COVID-19 demonstrate that the nature and magnitude of risk have changed. Due to COVID-19 and the climate challenge becoming simultaneous priorities, the post-pandemic recovery will be a crucial time for combating climate change. The positive effects of the COVID-19 pandemic lockdown on global environmental and planetary health are significant, despite the fact that they are not permanent and will not independently address climate change and environmental challenges as the world gradually returns to normalcy.

However, recognising that COVID-19 is first and foremost a global disaster, the pandemic may inspire future behavioural changes with positive environmental effects and serve as a practical lesson to the world that environmental degradation and climate change are still reversible and that it is possible to take drastic measures to achieve the United Nations' Sustainable Development Goals for the long-term environmental health of the planet. This includes initiatives such as the extensive use of clean energy.

COVID-19 demonstrated the systemic nature of risk by demonstrating how it not only affects certain components of our social and economic systems, but also constitutes a threat to their general functionality. However, not everyone in society has been harmed equally; those who are most susceptible to natural calamities have been hit the worst. The significance of a multisectoral, multihazard approach to disaster prevention, preparedness, response, and recovery planning has been emphasised. Systematically analysed, it is obvious that the risk is comprised of multiple components, including complex interdependencies between connected industries. Transnational risks with cascading effects across borders, regions, and system domains necessitating multi-level governance strategies; non-linear changes with potentially exponentially increasing impacts that are skewed in distribution and difficult to predict based on historical data or experience. Insufficient policy instruments that prevent institutional inertia and hasty legislative action while providing proper risk perception. This complexity impedes effective governance and necessitates a move from a single-hazards focus to a multi-hazards and systematic risk approach in emergency prevention, planning, response, and recovery.

The challenges of COVID-19 are deep and broad, but we have the opportunity to use these challenges to fast-track development of better frameworks for systemic risk management and contribute to building a better future.

12.4 Framework for Systemic Risk Management

Despite tremendous progress in this area, systemic risk governance requires additional maturation. Manifestations of compound, cascading, and systemic risks depict that the increasing and complex nature of risk is difficult to manage unless it is addressed from a systems approach. To this end, this study takes 'Resilience' as its central theme and provides the fundamental design principles and conceptual framework for Integrated Disaster Resilience (IDR) based on a 'Whole Systems Approach".

The proposed framework as depicted in Fig. 12.1, extends the GRAF 2020–2030 (UNDRR 2019) to highlight specific considerations required for assessing and managing risks. It leverages the 'Design-Manage-Evolve' methodology towards the holistic management and governance of the risk landscape and is guided by two key principles for integrating resilience into the program.

Whole Systems' Approach

The framework provides a comprehensive approach to creating resilience across systems and sectors, with an emphasis on lowering instability and unpredictability throughout the larger ecosystem and minimizing risk or possible damage to a person, group, community, or whole society.

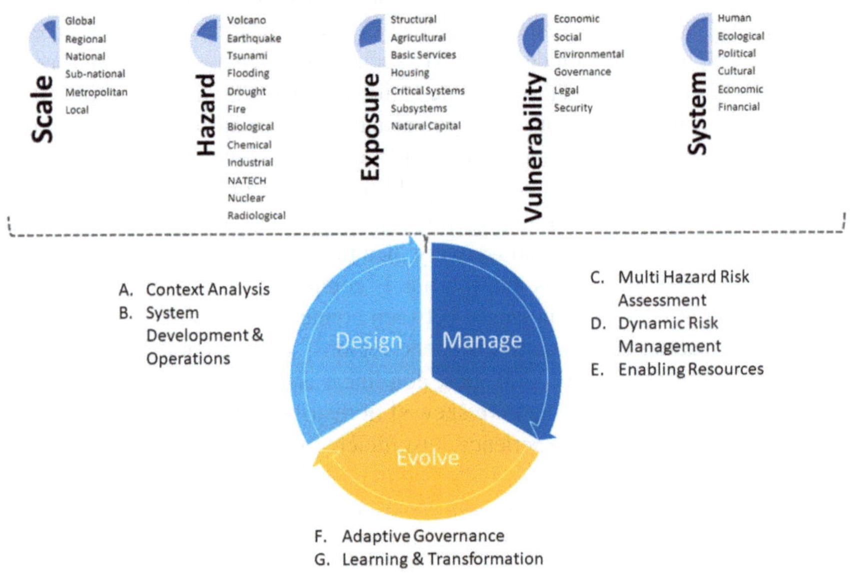

Fig. 12.1 Conceptual framework for integrated disaster resilience (IDR) based on a whole systems approach

Integrated Climate and Disaster Risk Management

The disaster risk reduction (DRR) and climate change adaptation (CCA) agendas place a premium on vulnerability reduction and resilience enhancement. Thus, it is critical to combine DRR and CCA to reap the benefits of risk reduction and adaptability concurrently. This research develops a conceptual framework that may aid in promotion of DRR and CCA integration by tying these two together to expedite risk reduction and adaptation efforts at all levels (global, regional, national, and local).

The framework is structured around three high-level dimensions that span the resilience building lifecycle. Describing a fluid relationship between key elements of the system and feed into each other.

12.4.1 Design Dimension

The 'Design' aspect emphasizes the importance of robust leadership, strategy formulation, and common objectives for sustainable decision-making in 'Resilience'. It involves integrating Disaster Risk Reduction (DRR) and Climate Change Adaptation (CCA) into developmental planning, fostering cooperative decision-making and goal alignment among essential systems, actors, and sectors. The phases of the design dimension start with Context Setting and Analysis, focusing on creating a comprehensive vision for systemic resilience and incorporating DRR and CCA into strategic planning and decision-making. This phase progresses to identifying and characterizing various systems in the disaster and climate risk environment, including their limits, sub-systems, dynamics, and main stakeholders, while also detailing their interconnections, potential conflict points, and areas of control, influence, and context.

Following the Context Setting and Analysis stage, the System Operations phase involves developing consistent DRR and CCA policies and integrating risk-informed planning across sectors, leading to the creation or modification of systems and infrastructures to be 'Resilient by Design'. This means they are recoverable and adaptable, including managing the overall system configuration, inter-system connections, and adding redundancies to reduce failure risks due to disruptions. Upon defining 'Resilient by Design' concepts, it's crucial to establish 'Key Performance Indicators' (KPIs) and 'Key Risk Indicators' (KRIs) for achieving common DRR and CCA objectives, while ensuring system performance standards. Quantifying resilience benefits is also vital for clarifying trade-offs between efficiency and resilience, guiding planning, implementation, and future investments. This dimension fosters 'Resilient Development and Design' through coherent policies, balanced power dynamics, and risk-aware decisions.

12.4.2 Manage Dimension

The 'Manage' dimension highlights the necessity of comprehensive risk evaluation and management through coordination among interdependent systems, preparing for, and responding to shocks and stresses with effective disaster response and resource allocation, including funding. The Manage Dimension stages include Risk Assessment, involving extensive risk evaluation to determine the full risk portfolio with single, compound, cascading, and systemic risks linked to underlying vulnerabilities. This is followed by identifying high-impact scenarios and their effects across different spatial and temporal scales. The Risk Management phase recommends recognizing connections between risk drivers and impacts, and focusing on strengthening interconnected systems. This involves a range of actions like institutional modeling, scenario analysis, stress testing, and creating dynamic disaster management plans featuring technological and nature-based solutions for functional and response diversity. Developing early warning systems and real-time decision-support tools using advanced technologies and detailed databases is suggested for proactive risk sensing and intelligence. Understanding risks through sense-making and creating trans-contextual relational information is vital for advancing systemic risk knowledge. The stage concludes with Enabling Resources, emphasizing the importance of accessible resources for effectively implementing the IDR framework, advocating for supportive regulations, innovative risk coverage through Disaster Risk Insurance (DRI), and forming inclusive DRR and CCA funds and incentive programs by public and private sectors.

12.4.3 Evolve Dimension

The 'Evolve' dimension emphasizes the necessity for governance systems that are adaptive, reflective, and transcend boundaries, coupled with the processes of learning and transformation within the constantly changing risk environment. This dimension stresses the importance of incorporating feedback mechanisms and ongoing learning as essential components of resilience, linking back to the 'Design' dimension to form a complete cycle. The phases within the Evolve Dimension include Adaptive Governance, which involves analyzing disaster and climate risks at various levels (global, regional, national, subnational, and community) through a governance approach that is multi-level and embraces multiple perspectives. This approach addresses systemic risks in varied geographical, social, and cultural settings and fosters a wide-ranging network of actors including markets, civil society, communities, and governments within adaptive governance. A key focus is also placed on proactive identification of future disaster and climate risks, acknowledging both progressive and defensive factors that contribute to resilience-enhancing strategies through the development of adaptive, absorptive, and transformative capacities.

Subsequently, the 'Learning and Transformation' stage of the framework concentrates on effective management of feedback loops and capacity building. It aims to promote self-organization and interconnectedness across different systems, providing pathways for the transition to Systemic Risk Governance, which is vital for effective governance in an evolving environment.

A mechanism for key decision-makers to better comprehend the systemic hazards affecting the disaster risk environment and act against them is provided by the conceptual framework that has been developed. However the planning and execution of this action will vary considerably depending on an organization's political, institutional, and cultural makeup as well as its willingness to address structural issues and lead change. A recent Systemic risk briefing note (International Science Council 2022) j published by the International Science Council (ISC), the United Nations Office for Disaster Risk Reduction (UNDRR) and the Risk Knowledge Action Network (Risk KAN) highlighted that to establish sustainable systems in the face of systemic risk, it is necessary to evaluate and handle the three connected and interacting 'spheres' of change—the practical, political, and human. While the practical realm is often dominated by 'technical responses' such as increasing knowledge and competence, fostering innovation, boosting management, and altering behavior. The political sphere comprises of systems, structures, and processes that reflect how society is structured, how systems work, and how systemic risks are handled or addressed. This is the area where decisions made or actions taken could endanger the interests and power relationships of current systems or foster resistance to change. Concludingly, the personal sphere includes both unique and shared beliefs, ideals, worldviews, and paradigms that influence attitudes and actions. These are the factors influencing both individual and societal perspectives on systems, as well as perceptions and attitudes toward systemic risks, which together often explain preferred tactics for practical reforms (O'Brien and Sygna 2013). It is improbable that attempting to address systemic risk in just one or two of the domains will result in sustainable change for risk reduction. All three realms must be taken into account when developing and communicating solutions for reducing systemic risks because they have an impact on how we perceive and categorise the system's numerous components. New approaches and policies will be required to identify and manage systemic risks, particularly disaster risks, in a hyperconnected future.

Adopting a systems-based approach to risk management has significant consequences for research and may pave the way for a new research agenda. The alternative paradigm of "Global Systems Science" examines if we need to know how to be robust to interconnected risks and how to be prepared for them, particularly in terms of economic stability, security, and prosperity, as well as public welfare.

Essentially, a multi-hazard, inter-disciplinary theoretical framework is necessary for catastrophe research, as well as a new institutional infrastructure, adaptive policy and regulatory architecture, new forms of legitimacy, dispersed capabilities, and global public interests, which will assist policymakers and regulators in making informed choices regarding systemic risks. Local and national, policies for disaster risk reduction should incorporate climate change scenarios and describe the circumstances necessary for transformative adaptation. Accelerated action on climate change will mitigate the effects of climate-related disasters and contribute to comprehensive risk management. This will also connect sustainability and resilience with the Sustainable Development Goals of the United Nations and Sustainable Cities and Communities. The objective is to establish a governance system that is adaptive, resilient, and inclusive with regard to systemic risks.

References

Academy TB (2021) The covid decade understanding the longterm societal impacts of COVID 19. https://www.thebritishacademy.ac.uk/documents/3238/COVID-decade-understanding-long-term-societal-impacts-COVID-19.pdf
al NP (2021) A framework for complex climate change risk assessment. One Earth
van Asselt MBA (2000) Perspectives on uncertainty and risk. Springer, Dordrecht. https://doi.org/10.1007/978-94-017-2583-5_10
Amnuaylojaroen T, Parasin N (n.d.) The association between COVID-19, air pollution, and climate change. https://www.frontiersin.org/articles/https://doi.org/10.3389/fpubh.2021.662499/full
Andreas Klinke OR (2002) A new approach to risk evaluation and management: risk-based, precaution-based, and discourse based strategies. Risk Annalysus.
Anne Tiernan LD (2019) A review of themes in disaster resilience literature and international practice since 2012. Policy Design and Practice
Arunabh Mitra RS (n.d.) Systemic risk from a disaster management perspective: a review of current research. https://www.sciencedirect.com/science/article/pii/S1462901122003690?dgcid=author
Axelsson J (2015) A systematic mapping of the research literature on system-of-systems engineering. In: 2015 10th system of systems engineering conference (SoSE). IEEE
Bak P (1996) How nature works: the science of self-organized criticality. Copernicus
Bansal S (2021) How COVID-19 advanced digital learning for lower-income populations. https://www.bcg.com/publications/2021/covid-19-advanced-digital-learning-for-lower-income-populations
Castellani B (2018) Map of complexity sciences. https://www.art-sciencefactory.com/complexity-map_feb09.html
Cavallo A (2014) Integrating disaster preparedness and resilience: a complex approach using system of systems. Australian J Emergency Manage
Centeno MA (2021) Global systemic risk. Coursera, Coursera, MOOC
Chakraborty BB (2021) Positive effects of COVID-19 lockdown on river water quality: evidence from River Damodar, India. https://www.nature.com/articles/s41598-021-99689-9
Christmas Uchiyama NI (2021) Assessing contribution to the Sendai framework: case study of climate adaptation and disaster risk reduction projects across sectors in Asia-Pacific (2015–2020). Progress in Disaster Science
Complex Systems Journal (n.d.) https://www.complex-systems.com/
Damon Centola MW (2007) Complex contagions and the weakness of long ties. Am J Sociol
Deborah SK, Thomas SJ (2020) The CHASMS conceptual model of cascading disasters and social vulnerability: the COVID-19 case example. Int J Disaster Risk Reduction
Dijkman M (2010) A framework for assessing systemic risk. RePEc (Research Papers in Economics)
Editor (2021) The long-term environmental implications of COVID-19. https://www.oecd.org/coronavirus/policy-responses/the-long-term-environmental-implications-of-covid-19-4b7a9937/
Editor (2022) COVID-19 drives global surge in use of digital payments. https://www.worldbank.org/en/news/press-release/2022/06/29/covid-19-drives-global-surge-in-use-of-digital-payments

Editor (2022) World development report 2022-Chapter 1. The economic impacts of the COVID-19 crisis. https://www.worldbank.org/en/publication/wdr2022/brief/chapter-1-introduction-the-economic-impacts-of-the-covid-19-crisis
Editor (n.d.) Retrieved from International Risk Governance Council. https://irgc.org/risk-governance/systemic-risks/guidelines-governance-systemic-risks-context-transitions/
Editor (n.d.) Global Risks Report 2021. https://www.weforum.org/reports/the-global-risks-report-2021/
Editor (n.d.) Chapter 1. The economic impacts of the COVID-19 crisis. Retrieved from World Development Report 2022. https://www.worldbank.org/en/publication/wdr2022/brief/chapter-1-introduction-the-economic-impacts-of-the-covid-19-crisis?
Engel H, Hamilton A, Nauclér T, Fine D, Pinner D, Rogers M, Leger S (n.d.) How a post-pandemic stimulus can both create jobs and help the climate. https://www.mckinsey.com/capabilities/sustainability/our-insights/how-a-post-pandemic-stimulus-can-both-create-jobs-and-help-the-climate
Galaz V (2014) Anthropocene risks: social scientists need to step up to the challenge. Guardian
Gerald Marwell PO (1993) The critical mass in collective action. Cambridge University Press
Granovetter M (1978) Threshold models of collective behavior. Am J Sociol
HannahSnyder (2019) Literature review as a research methodology: an overview and guidelines. J Bus Res
Henry Ngenyam Bang LS (2019) Disaster risk reduction in cameroon: are contemporary disaster management frameworks accommodating the sendai framework agenda 2030? Int J Disaster Risk Sci
Helbing D (2013) Globally networked risks and how to respond. Nature
Holling CS (2001) Understanding the complexity of economic, ecological, and social systems. Ecosystems
Hong Chuan Loh IL (n.d.) Positive global environmental impacts of the COVID-19 pandemic lockdown: a review. https://link.springer.com/article/https://doi.org/10.1007/s10708-021-10475-6
Hui-Min LI, Xue-Chun WA (2021) Understanding systemic risk induced by climate change. Adv Clim Change Res
Huizen J (n.d.) How COVID-19 has changed the face of the natural world. https://www.medicalnewstoday.com/articles/how-covid-19-has-changed-the-face-of-the-natural-world
Hung LS (2003) The SARS epidemic in Hong Kong: what lessons have we learned? J R Soc Med
Ian Goldin MM (2015) The butterfly defect: how globalization creates systemic risks, and what to do about it? Princeton University Press
International Science Council (2022) Retrieved from briefing note on systemic risk : s://council.science/publications/briefing-systemic-risk/
IPCC (2014) IPCC, 2014: Climate Change 2014: Synthesis Report. IPCC.
Jason Channell EC (2021) Systemic risk: systemic solutions for an increasingly interconnected world. Centre for Risk Studies at Cambridge University, Citi Global Perspectives & Solutions
Kauffman SA (1993) The origins of order: self-organization and selection in evolution. Oxford University Press
Keating A, Keating C (2014) Operationalizing resilience against natural disaster risk: opportunities, barriers, and a way forward. Zurich Flood Resilience Alliance
Hagelsteen M, Becker P (n.d.) Retrieved from Systemic problems of capacity development for disaster risk reduction in a complex, uncertain, dynamic, and ambiguous world. https://www.researchgate.net/publication/331380101_Systemic_problems_of_capacity_development_for_disaster_risk_reduction_in_a_complex_uncertain_dynamic_and_ambiguous_world
Kugler M, Viollaz M, Duque D, Gaddis I, Newhouse D, Palacios-Lopez A, Weber M (2021) How did the COVID-19 crisis affect different types of workers in the developing world? https://openknowledge.worldbank.org/handle/10986/35823
Kumar ASP (n.d.) Impact of COVID-19 on greenhouse gases emissions: a critical review. https://www.ncbi.nlm.nih.gov/pmc/articles/PMC8445775/

Kumar A, Dubey A, Malla MA (n.d.) With corona outbreak: nature started hitting the reset button globally. https://www.frontiersin.org/articles/https://doi.org/10.3389/fpubh.2020.569353/full
Kwasinski AJ (2017) Further development of a conceptual framework for assessing resilience at the community scale. NIST
Lloyd's (2021) Retrieved from Lloyd's. https://www.lloyds.com/news-and-insights/futureset/systemic-risk
Luhmann N (1995) The paradoxy of observing systems. Cultural Critique
Lund S, Madgavkar A, Manyika J, Smit S, Robinson O, Meaney M, Ellingrud K (n.d.) The future of work after COVID-19. https://www.mckinsey.com/featured-insights/future-of-work/the-future-of-work-after-covid-19?fbclid=IwAR1dp9Yh3eK_kSTmRycaAfOf3Gt1oshKvCX-8
Macy DC (2007) Complex contagions and the weakness of long ties. Am J Sociol
Magnus Hagelsteen PB (2019) Systemic problems of capacity development for disaster risk reduction in a complex, uncertain, dynamic, and ambiguous world. Int J Disaster Risk Reduction
Mari Nicola ZA-J (2020) The socio-economic implications of the coronavirus pandemic (COVID-19): a review. https://www.ncbi.nlm.nih.gov/pmc/articles/PMC7162753/
Mark Wever NW (2021) Towards a transdisciplinary approach to systemic risk detection. In: The 28th ISTE international conference on transdisciplinary engineering (TE2021) at: university of bath, england
May H (2011) Systemic risk in banking ecosystems. Nature
Miguel A, Centeno MN (2015) The emergence of global systemic risk. Ann Rev Sociol
Mitchell M (2009) Complexity: a guided tour. OUP USA
Mearsheimer (2009) The case for a ukranian nuclear deterrent. Foreign affairs
Mousazadeh MPB (n.d.) Positive environmental effects of the coronavirus 2020 episode: a review. https://www.ncbi.nlm.nih.gov/pmc/articles/PMC7859094/
Muduli PK (n.d.) Water quality assessment of the Ganges River during COVID-19 lockdown. https://doi.org/10.1007/s13762-021-03245-x
O'Brien K, Sygna L (2013) Responding to climate change: the three spheres of transformation1. In: Changing climate international conference. Oslo
OECD (2002) Emerging risks in the 21st century: an agenda for action
Ortwin Renna KL (2017) Things are different today: the challenge of global systemic risks. J Risk Res
Patrick W, Keys VG (2019) Anthropocene risk. Nat Sustain
Page JH (2007) Complex adaptive systems: an introduction to computational models of social life. Princeton University Press
Perrow C (1984) Normal accidents. Princeton University Press
Petri Uusikylä PT (n.d.) Disaster management as a complex system: building resilience with new systemic tools of analysis. In: Society as an interaction space, a systemic approach, 2020
Pia-Johanna Emma Schweizer OR (2019) Governance of systemic risks for disaster prevention and mitigation. Disaster Prevention and Manage
Pia-Johanna Emma Schweizer OR (2019) Governance of Systemic Risks for Disaster Prevention and Mitigation. In: Global assessment report on disaster risk reduction 2019.
Polinpapilinho F, Katina CB (2019) A governance perspective for system-of-systems. Systems, MDPI
Renn O, Laubichler M, Schanze J, Scholz R, Schweizer P (2020) Systemic risks from different perspectives. Risk Anal
Renn AK (2006) Systemic risks as challenge for policy making in risk governance. Forum: Qual Soc Res
Renn O (2016) Systemic risks: the new kid on the block. Environ: Sci Policy Sustain Dev
Rittel W (1973) Dilemmas in a general theory of planning. Policy Sci
Riyanti Djalante CH (2011) Adaptive governance and managing resilience to natural hazards. Int J Disaster Risk Sci
Riyanti Djalante CH (2013) Pathways for adaptive and integrated disaster resilience. Nat Hazards

Riyanti Djalantea SL (2019) Governing complexities and its implication on the Sendai framework for disaster risk reduction priority 2 on governance. Progr Disaster Sci
Scheffer M (2009) Critical transitions in nature and society. Princeton University Press
Scheffer M (2010) Foreseeing tipping points. Nature
Schweizer P (2019) Governance of systemic risks for disaster prevention and mitigation. https://www.preventionweb.net/files/66695_f431finalschweizergovernanceofsyste.pdf
Scira Menonia MP (2020) A systemic approach for dealing with chained damages triggered by natural hazards in complex human settlements. Int J Disaster Risk Reduction
SFDRR (n.d.) https://www.undrr.org/publication/sendai-framework-disaster-risk-reduction-2015-2030
Md. Shahab Uddin JK (2019) Systems thinking approach for resilient critical infrastructures in urban disaster management and sustainable development. In: Resilient structures and infrastructure. Springer
Shaw A (2021) A systems thinking approach to risk reduction and mitigation for improving disaster management. Chan School of Public Health, Harvard T.H
Shroff A (2021) A systems thinking approach to risk reduction and mitigation for improving disaster management. Chan School of Public Health, Harvard T.H
Simpson PN (n.d.) A framework for complex climate change risk assessment
Stefan Hochrainer SC (n.d.) Enhancing resilience of systems to individual and systemic risk: steps toward an integrative framework. Int J Disaster Risk Reduction, 2020
Stengers IP (1984) Order out of chaos: man's new dialogue with nature. The Br J History of Sci
Susan Lund AM (2021) The future of work after COVID-19. https://www.mckinsey.com/featured-insights/future-of-work/the-future-of-work-after-covid-19
Terje Aven OR (2019) Some foundational issues related to risk governance and different types of risks. J Risk Res
University UN (2021) Disaster risks in an interconnected world. UN
UNDP (2021) Social construction of systemic risk towards an actionable framework for risk governance
UNDRR (2019) UNDRR
UNDRR (2020) The human cost of disasters
UNDRR (n.d.) Prevention web. https://www.preventionweb.net/understanding-disaster-risk/risk-drivers/climate-change
University C (2021) Risk prevention, Cambridge judge business school. https://www.jbs.cam.ac.uk/insight/2021/risk-prevention/
University C (2021) Risk, prevention, cambridge judge business school. https://www.jbs.cam.ac.uk/insight/2021/risk-prevention/
UNU-EHS U (2022) Understanding and managing cascading and systemic risks. Geneva
Watts DJ (2004) The "New" science of networks. Ann Rev Sociol
WEF (2014) The Global Risks Report 2014
WEF (2017) The Global Risks Report 2017. WEF
WEF (2021)
Weiner N (n.d.) Cybernetics. 2019
Yong-kyunKim WC-G-L (2022) A strategic response map for cascading pandemics: Lessons learned from the response to COVID-19 in the Republic of Korea. Progr Disaster Sci
Yunfeng Shang HL (2021) Effects of pandemic outbreak on economies: evidence from business history context. https://www.frontiersin.org/articles/https://doi.org/10.3389/fpubh.2021.632043/full

Part III
The All-hazards Approach and it d Mechanism and System

Chapter 13
Challenges in Promoting All-Hazard Approach in Japan: Institutional Arrangements for Managing Cascading Effects

Mikio Ishiwatari and Hideki Hirai

Abstract Societal and climate change are increasing disaster risks. As the world becomes more interconnected, disasters disrupt socioeconomic activities on a global scale more severely. The COVID-19 pandemic became a global disaster on a scale that transcended ecological regions, national borders, economies, and societies. The pandemic has impacted a wide range of social and economic sectors and shut down core businesses, leading to a worldwide recession. Countries need to develop capacity in terms of disaster risk reduction to cope with complexity, novelty, ambiguity, and uncertainty that characterize modern disasters. An all-hazard approach could respond to various hazards in an integrated manner. This study investigates the issues of taking such an approach through examining discussions over policies and institutions in Japan. In-depth interviews were conducted with government officers and researchers in September 2022. Government reports and academic papers were reviewed. Japan has improved the policies and institutions that respond to emerging hazards. The Great East Japan Earthquake and Tsunami in 2011 caused cascading effects on sensitive facilities and became a highly complex phenomenon. The tsunami damaged the Fukushima Daiichi nuclear power station, leading to disruption of energy supply, environmental deterioration, and threats to public health. Direct damage to major industries affected industrial production through the dense networks of supply chains around the world. To respond to the COVID-19 pandemic, Japan, like other countries, is improving the systems of managing epidemics. The country has established no comprehensive legislation of emergency management but practically manages this

M. Ishiwatari (✉)
The University of Tokyo, Kashiwa City, Chiba, Japan
e-mail: Ishiwatari.Mikio@jica.go.jp

H. Hirai
Water Resources Environment Center, Chiyodaku, Tokyo, Japan

T. Izumi et al. (eds.), *All-Hazards Approach*, Disaster Risk Reduction,
https://doi.org/10.1007/978-981-97-1860-3_13

based on the segregated laws of managing disasters, nuclear accidents, influenza and epidemic, and armed conflicts. While the Cabinet Secretariat leads the responses to emergencies affecting human lives and assets, four offices in the government are mandated separately to manage these events. Line ministries and agencies have accumulated knowledge on disaster management by disaster type and trained staff.

Keywords Cascading disaster · Cascading effects · COVID-19 · Great East Japan Earthquake and Tsunami

13.1 Introduction

To promote sustainable development, approaches to managing the cascading effects caused by disasters need to be established. Societal changes, such as urbanization and modernization, as well as climate change are increasing disaster risks. As the world becomes more interconnected, disasters disrupt socioeconomic activities through supply chains on a global scale more severely. The development of information and communication technology (ICT) and promotion of transport and free trade have created an extensive network of production, trade, and investment worldwide (Alexander 2018; Chen et al. 2019; Galanakis et al. 2021).

The Great East Japan Earthquake and Tsunami (GEJE) in 2011 caused cascading effects on the sensitive facilities of nuclear power plants, oil refineries, highways, ports, and hospitals, and became a highly complex phenomenon of a cascading disaster. The COVID-19 pandemic became a global disaster that transcends ecological regions, national borders, economies, and societies (Alcántara-Ayala et al. 2021). The pandemic has impacted various social and economic sectors and shut down core businesses, leading to a worldwide recession and growth diversion (DeWit et al. 2020; Yu et al. 2022). Disaster management has become complicated because disaster management offices need to manage the spread of COVID-19 during emergencies (Inoue et al. 2022; Ishiwatari et al. 2020). Climate change is expected to cause more severe cascading effects as a result of climate disasters including flooding, landslides, droughts, and heatwaves (Hoeppe 2016; Kjekstad and Highland 2009; Ministry of Environment 2020).

To manage increasing risks, the OECD (2018) recommends that each country should develop an all-hazard approach, which can manage various hazards in an integrated manner. This approach should cope with the complexity, novelty, ambiguity, and uncertainty that characterize modern disasters.

This study examines the issues of adopting an all-hazards approach by identifying changes in policies and institutions in Japan. In-depth interviews were conducted with government officers and researchers in Tokyo and Sendai in September and Washington DC in December 2022. Government documents and research papers were reviewed. This study has been prepared as a part of the research project "Demand estimate on social and disaster prevention infrastructure in Asia" conducted by the JICA Ogata Research Institute, Tokyo, Japan.

13.2 All-Hazard Approach to Manage Cascading Effects

Cascading effects caused by disasters "*generates a sequence of events in human subsystems that result in physical, social, or economic disruption*" (Pescaroli and Alexander 2015). The effects cover various sectors of health, industrial and economic activities, and people's daily lives by damaging infrastructure. Climate change is increasing these effects. The impacts caused by climate change directly damage human lives, agricultural products, houses, and industry. Disasters deteriorated by climate change affect crucial infrastructure and commercial facilities as well as disrupt the public services of electricity, communication, water supply, sewerage system, transportation, and solid waste management (Fig. 13.1; Ministry of Environment 2020).

Understanding complex risks by identifying a wide range of disaster scenarios is a fundamental effort to manage risks (OECD 2018). Risk governance should be arranged to decide the roles and responsibilities of organizations concerned (Kishimoto 2015). The Science Council of Japan (2020) recommends that government organizations and experts prepare for large-scale disasters by considering the uncertainty of scientific knowledge to avoid "unexpected" disasters. The scientific community needs to communicate with government organizations and local communities that disaster predictions are not certain and have a range. Although cascading disasters cannot be fully managed due to their high complexity and cross-scale dynamics, each country should examine potential vulnerabilities and cascading paths and prepare for disasters in advance (Pescaroli and Alexander 2016). They should allocate physical and human resources by coordinating organizations concerned (Okamura 2012).

Since anticipating all effects caused by disasters is difficult, the US and European countries have strengthened their all-hazard approaches (Nagata 2022). Which hazards cause disasters is not important to respond to emergencies. Regardless of causes, a society needs to respond to all emergencies. Mutai and Leo (2004) argue that a society can respond to an emergency better by adopting an all-hazard approach than approaches developed for single events.

In the US, the Federal Emergency Management Agency (FEMA) is adopting the aforementioned approach. The FEMA is the leading agency of emergency management in the federal government in terms of coordinating concerned organizations. Once a US president declares a state of emergency, the FEMA manages all resources of disaster management in the Federal Government. The FEMA formulated the National Response Framework, which are categorized and prioritized by the disaster response mechanism and covers 15 response areas (Mutai and Leo 2004). The framework adopted a multi-agency system to identify coordinators and primary support organizations. Disaster management organizations apply the Incident Command System, which is a standardized chain of command under one system throughout the US (Okamura 2012). The Pandemic and All-Hazards Preparedness Act of 2006 mandated the Department of Health and Human Services to lead preparing for public health emergencies. Further, the Pandemic and All-Hazards Preparedness

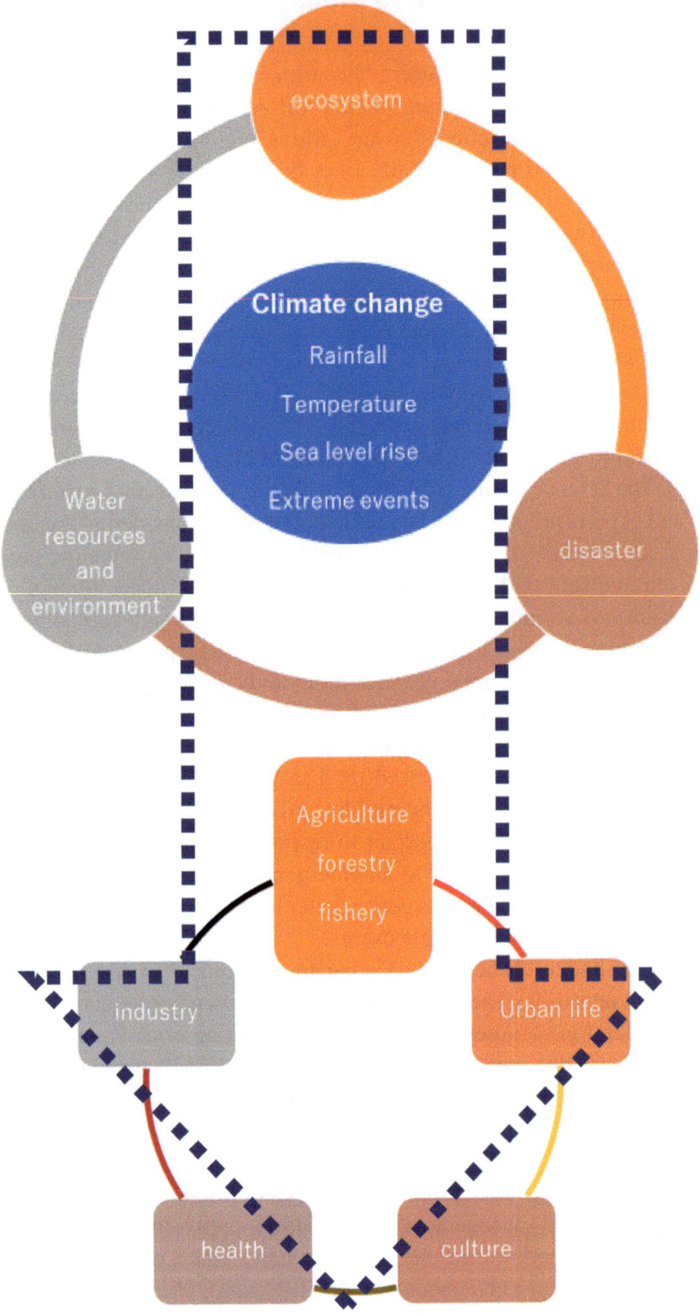

Fig. 13.1 Cascading effects by climate change. *Source* Modified from Ministry of Environment (2020)

Reauthorization Act of 2013 aims at enhancing response capacity and research and development in the medical area (Kato and Yamaguchi 2019).

In the United Kingdom, the Contingencies Secretariat in the Cabinet Office is responsible for emergency management. In normal times, national risk assessment is conducted, and the National Risk Register, which shows risk scales by hazard, is presented (Kishimoto 2015). The register encourages the private sector, local communities, and ordinary citizens to prepare for the emergencies of natural disasters, cyberattacks, terrorism, and epidemics.

All European countries have established emergency management offices to coordinate multiple organizations in their governments. Most offices function as the secretariat of decision-making in emergency response to coordinate government ministries and agencies (Nagata 2022).

While Japan has improved its policies and institutions to respond to emerging hazards, it does not explicitly adopt the all-hazard approach (Ishiwatari 2021). The country experienced the GEJE, which damaged sensitive facilities and caused cascading effects. The tsunami damaged the Fukushima Daiichi nuclear power station, leading to disruption of energy supply, environmental deterioration, and threats to public health. Direct damage to major industries affected industrial production through the dense networks of supply chains around the world (Ranghieri and Ishiwatari, 2014). Leo (2012) highlighted various issues of unrealistic and non-comprehensive response plans, lack of uniform incident management systems, underutilization of volunteers and nongovernment organizations, one way and top-down communication systems, no shelter management plan, unmanaged relief goods delivery, over-reliance on Self-Defense Force, limited human resources in disaster management offices, and no feedback systems to use lessons learned from disasters. Following the disaster, the country improved the management systems of nuclear accidents and established the Nuclear Emergency Response Headquarters, chaired by the Prime Minister.

Similar to other countries, Japan has struggled to respond to the COVID-19 pandemic (Inoue et al. 2022). The government has expanded its offices and institutions. In January and February 2020, the government established a ministerial council and panel of experts, respectively. Furthermore, it established a new Office for Novel Coronavirus Disease Control in March (Ishiwatari 2020). Currently, the country is improving the system further based on lessons learned from the pandemic.

While coordination among concerned offices has been strengthened, Japan has yet to implement a comprehensive all-hazard approach. Instead, individual institutions have been established to manage pandemics and nuclear accidents separately.

13.3 Mechanisms of Managing Disasters in Japan

This section reviews disaster management mechanisms in Japan from the perspective of the all-hazard approach. The issues and new systems of managing epidemics following the COVID-19 pandemic, the current mechanisms of managing emergencies, and the all-hazard approaches of the private sector are examined.

13.3.1 Establishing Management Mechanisms of Infectious Diseases Through Responding to the COVID-19 Pandemic

As the government needed to respond to the cascading effects caused by the COVID-19 pandemic, multiple groups and committees of experts were established to formulate policies to control the pandemic (Tashiro and Shaw 2020). However, the processes of policymaking were not clear. While the government applied most of the over 70 policies recommended by experts, it did not explain the reason for not applying some of these (Nikkei 2021).

Various systems and applications were released but were not sufficiently well designed to lead to behavior change through daily use (Ide 2021). In an attempt to control the spread of COVID-19, ordinary people have been forced to use a contact tracking app. However, people's decision to use the app is influenced by social factors, such as social stigmatization and lack of trust in experts. Japanese people are concerned regarding relationships with others in society (Jamieson et al. 2021). Medical experts and IT specialists should work together to improve the usability of these systems from the user's perspective and deepen data collaboration. Delivering a message based on the values of various groups was a challenge. The government should implement policies together with communities based on their concerns and perceptions, rather than from the top-down, "ruler's perspective" (Nikkei 2021).

The balance between economic and pandemic control policies was unclear. The government subsidized and promoted tourism during the pandemic through the "Go To Travel" campaign, which had the potential to spread the virus throughout the country (Reich 2020).

Japan will establish a new agency in the Cabinet Secretariat, the Cabinet Integrated Agency of Infectious Disease Crisis Management, to oversee the response to infectious disease crises and serve as the government's control tower. In the event of an outbreak of infectious disease, the agency is responsible for the initial response and coordination among relevant ministries and agencies by collecting and analyzing information and organizing meetings. The Prime Minister supervises the response and maximizes the use of human resources within the government. A list of staff to be called upon in the event of an emergency will be prepared in advance. A new specialized center on infectious diseases to be created in the Health Ministry, which is known as the Japanese version of the Center for Disease Control and Prevention of the US, will work closely together from normal times, for example, by receiving information on high-quality scientific evidence on infectious diseases.

In normal times, the agency will conduct drills of infectious disease emergency management, raise public awareness, and examine and improve the preparedness of each ministry and agency.

13.3.2 Government Systems Managing Hazards

The Emergency Preparedness and Crisis Management Office of the Cabinet Secretariat (CMO-CS) leads the response to any large-scale disasters and crises by coordinating government organizations at the national level. The Headquarters for Emergency Disaster Control, which is chaired by the Prime Minister, is established once a large-scale disaster happens. GEJE is the first case in which this was established. For other disasters, the Headquarters for Extraordinary Disaster Control, which is chaired by the Minister for Disaster Management, is established. In an emergency, an emergency response team that consists of the senior officials of relevant ministries and agencies gather at the Prime Minister's Office to collect information on disasters and conduct initial response measures. These mechanisms have been developed based on lessons learned from disasters, mainly following the Great Hanshin-Awaji (Kobe) Earthquake in 1995; the response to the earthquake was delayed and not well coordinated.

There are four specialized offices and multiple laws to manage natural disasters, nuclear accidents, epidemics, and civil protection from armed conflicts (Table 13.1). The Disaster Management Office in the Cabinet Office (DMO-CO) is responsible for planning basic policies on management of natural disasters and response to large-scale disasters through coordinating organizations at the national level. This office was also established following the Kobe Earthquake to strengthen institutions at the national level (Ishiwatari et al. 2021).

Table 13.1 Current institutional arrangement at the national level

Hazards	Law	Organization in charge
Natural disasters, fire, and accidents	Disaster management basic law	Disaster management office of cabinet office
Nuclear accidents	Law on countermeasures against nuclear accident	Nuclear regulatory commission
Epidemic	Influenza law	Cabinet integrated agency of infectious disease crisis management, cabinet secretariat
Armed conflicts	Civil protection law	Crisis management office, cabinet secretariat

Source Modified from the Council of Vice-Ministers concerned with examining government emergency management institutions (2015)

The Nuclear Disaster Prevention Office was established in the Cabinet Office in 2014 to enhance nuclear disaster management systems at the local level, and to serve as the secretariat for the Nuclear Emergency Preparedness Council and the Nuclear Emergency Response Headquarters. Following the Fukushima nuclear accident in 2011, the system was improved. The Nuclear Disaster Prevention Office coordinates relevant ministries, agencies, and local governments on nuclear disaster management in both normal times and emergency situations. To manage both nuclear accidents and natural disasters, in 2015, information collection, decision-making, and instructions and coordination were integrated between two headquarters of Emergency Disaster Control and Nuclear Emergence Response.

The civil protection law was enacted in 2004 to protect the peace and independence of Japan and to maintain the security of the nation and its people in the event of an armed attack against the country. The government established policies to protect the public, issue warnings, and direct evacuation measures. The country's concern stemmed from the 9/11 attacks and missile development by a neighboring country, which has led Japan to be vigilant about potential threats (Fire and disaster management Agency (FDMA) 2017). The CMO-CS leads coordination among ministries and local governments.

13.3.3 Private Sector

The private sector in Japan started adopting all-hazard approaches to continue their businesses in times of disaster. Since Japanese companies faced difficulties in operating during the COVID-19 pandemic, they have understood the necessity of strengthening the measures of continuing their businesses during disasters.

Japan Business Federation and KEIDANREN (2021) suggested companies adopt all-hazard approaches for their business continuity plans (BCP). The federation is a comprehensive economic organization with a membership of some 1,500 companies covering major Japanese companies, and provides policy advice to the Japanese government and society from the perspective of the private sector. The federation stresses the importance of shifting from a response based on individual disasters such as pandemics and natural disasters to an all-hazards response. Companies should prepare for "resulting events," regardless of the cause of the emergency, to continue their operations even in the event of a complex disaster, such as the simultaneous occurrence of an infectious disease epidemic and a natural disaster. Companies need to prioritize works to be continued or recovered during disasters, and to allocate and develop resources to continue the core works.

The Japanese government did not originally recommend that companies adopt an all-hazard approach for formulating BCPs, instead to prepare for earthquakes as the first step. The first version of the 2005 guideline (Central Disaster Management Council 2005) recommends that companies formulate plans to prepare for earthquakes, which have high risks and foreign countries are concerned for Japanese business continuity. The guidelines suggest setting goals against earthquakes at each important business operation site and considering specific measures necessary for business continuity. This is because the Japanese can understand the response to earthquake disasters more easily than other disasters, since earthquakes frequently occur throughout history. The government has revised BCP guidelines to expand targeted disasters. The revised guidelines suggest considering a scenario that is slightly worse than the most likely scenario, not the worst-case scenario.

The third version, which was formulated in 2013 following the GEJE in 2011, expanded targeted disasters to all natural disasters and became applicable to man-made disasters. The Miyagi Prefecture Government (2014) formulated BCP guidelines that cover a wide range of natural disasters, such as tsunamis, earthquakes, and all other hazards. This guideline incorporates lessons learned from the GEJE, such as the importance of the leadership and flexibility, targeting all business operations, regular training and exercises, and a stepwise approach.

13.4 Institutional Arrangements Proposed for All-Hazard Approach

Various organizations and researchers have proposed or examined institutional arrangements for managing disasters at the national level (Fig. 13.2). Recently, arrangements have discussed to consider a follow-up organization of the Recovery Agency, which was established in 2012 for ten years to promote recovery activities from GEJE in 2011(Mori 2019). Recommendations focus on strengthening the national functions of disaster management by continuing and expanding the roles of the Recovery Agency.

A research group recommended establishing a disaster management ministry to coordinate policies and enhance the capacity of disaster management by securing specialists (Consultation Group for Japanese Disaster Management 2017). This group

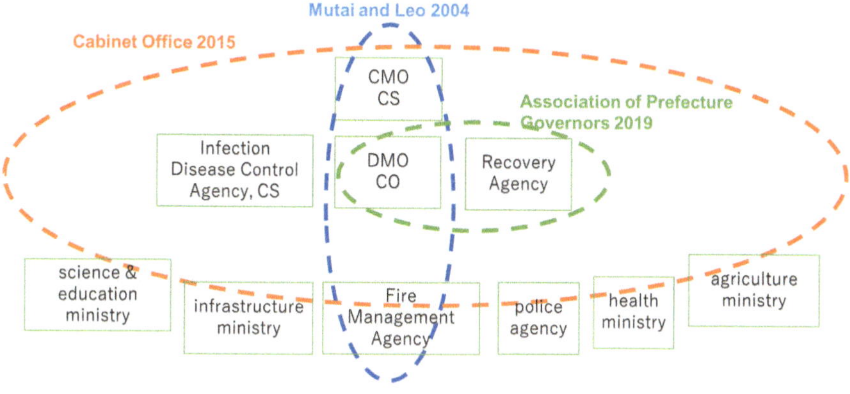

Fig. 13.2 Concepts for institutional arrangements. *Source* Compiled by the authors

is supported by prefectural governors in the Kansai Region. Currently, most staff of DMO-CO are seconded from other ministries and transferred every two to three years. With specialists trained, a new ministry is expected to prepare for large-scale disasters through raising public awareness, mainstreaming disaster management, managing disaster information, supporting local governments, managing knowledge, and back-upping capital functions. The association consisted of all prefecture governors requested the national government create a disaster management ministry by merging the Recovery Agency with DMO-CO (Association of Prefecture Governors 2019).

Mutai and Leo (2004) argue that the Japanese mechanism of emergency management lacks the headquarter function, training opportunities, strategies, and planning, while each staff member has high capacity. In addition, the roles and responsibilities of each staff member are not clearly stipulated. They recommend that Japan should adopt the all-hazard approach instead of the current hazard-specific approach and that Japan needs the leading agency to make decisions on emergency response by coordinating all organizations concerned. They further argue that the three organizations of the Fire Management Agency, DMO-CO, and CMO-CS should be merged to establish the leading agency. Similarly, to strengthen emergency management in public health, Kato and Yamaguchi (2019) recommend establishing a FEMA-type office in Japan. They argue that to adopt an all-hazard approach to manage cascading disasters, the emergency management of public health needs to be improved.

Kishimoto (2015) argues that Japanese approaches do not cover risk governance in the resilience strategy. He suggests that the country should start risk assessment and mapping covering all hazards, including low-probability ones.

The Japanese government believes that integrating the concerned organizations has no advantage (Council of Vice-ministers concerned with examining emergency management institutions of the government, 2015). The government evaluated that the current mechanism of emergency management led by the Cabinet Secretariat and supported by line ministries and agencies functions could be justified, and does not need to establish a leading agency.

13.5 Considerations

Various papers and organizations recommend that a FEMA-type of organization should be established to adopt an all-hazard approach in Japan. However, given the current mechanisms led by the Cabinet Secretariat functions, modifying institutions needs careful consideration (Fig. 13.3; Council of Vice-ministers concerned with examining emergency management institutions of the government, 2015). This section discusses factors affecting institutional arrangements.

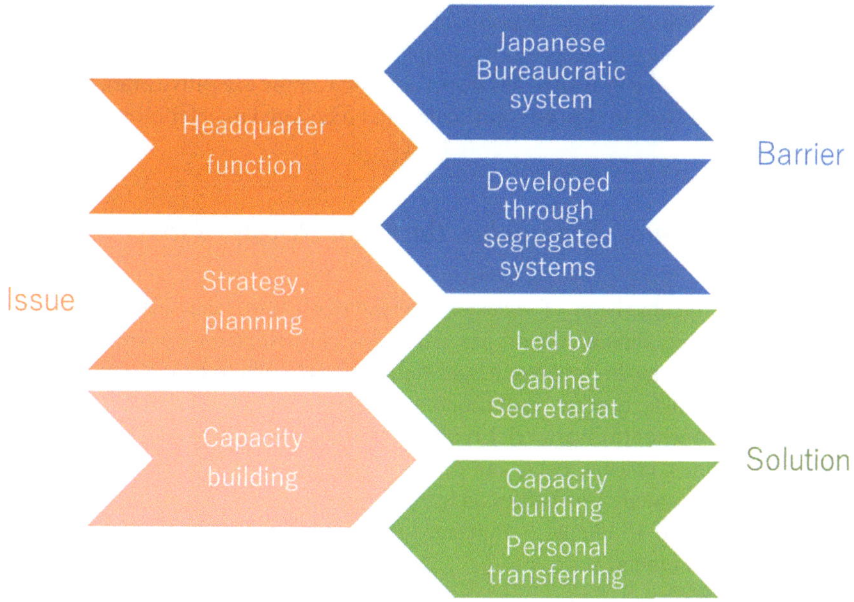

Fig. 13.3 Issue, barrier, and solution of all-hazard approach in Japan. *Source* Compiled by the authors

13.5.1 Political and Bureaucratic Systems

The US systems that clearly mandate staff responsibility appear ideal, but transplanting the same system would not fit with the Japanese bureaucratic system. Creating a new organization needs to be in line with Japanese political structures and bureaucratic mechanisms (Machidori 2015). Unclear mandates in the Japanese system are not limited to disaster management. The system does not explicitly authorize the responsibilities of positions. This may be considered weak but can also be considered as a flexible system to supplement a position unworked by other staff (Hata 2022).

Another issue is differences between the Japanese cabinet and US presidential systems. Japan's cabinet system is characterized by independent ministries and the weak authority of the Prime Minister (Sasaguchi 2014). The Japanese system needs coordination mechanisms even when the Prime Minister leads an emergency response. Under the US presidential system, the President serves as both the head of state and head of government and has strong authority over executive power to manage various departments and agencies. The President has the authority to appoint and remove top officials, and to set policies and priorities for the government.

This fundamental difference in the structure of the two systems has significant implications for disaster management. In Japan, the lack of direct authority over the various ministries makes it difficult to establish effective disaster management mechanisms that work within the political and bureaucratic structures of the country. This challenge cannot be resolved by simply creating a new leading institution. In contrast, in the US, the President has direct control over the executive branch and can direct response efforts in a more coordinated and integrated manner through FEMA.

13.5.2 Segmented Approaches Based on Disaster Types

The Japanese mechanisms of disaster management have been established in a segmented way based on disaster types. There has been limited focus on developing an integrated or all-hazard approach, and staff have not been trained to work in such a manner. Specialists, experts, practitioners, and researchers have been trained separately, and approaches to disaster risk reduction have evolved in silos for floods, earthquakes, landslides, coastal, fires, and others. The Basic Plan for Disaster Management, a national strategy, consists of an introduction, a chapter on common measures for all hazards, and 13 chapters on hazard-specific measures for floods, tsunamis, earthquakes, and others.

13.5.3 Lead by Cabinet Secretariat

The Cabinet Secretariat can use the professional knowledge and expertise accumulated by line ministries. For example, the Ministry of Land, Infrastructure, Transport, and Tourism (MLIT) is responsible for flood protection, and the staff of river engineers have been trained and engaged in flood protection for policymaking at the headquarters and actual works in the field. When flood disasters occur, the ministry dispatches key staff and provides necessary information to the CMO-CS.

It is unrealistic to separate staff from the MLIT and transfer them to the lead agency for disaster management. This is because they do not always engage in emergencies. These staff members manage flood protection facilities during normal times and are engaged in emergency response once a disaster occurs. Similarly, the Ministry of Health, Labour and Welfare is responsible for managing epidemic disasters, the Ministry of Defense is responsible for public protection from armed conflicts, and the Ministry of Education, Culture, Sports, Science, and Technology is in charge of monitoring and predicting disasters.

Even though the US system is regarded as the model of the all-hazard approach, government organizations are not totally integrated. The US Army Corps of Engineers is in charge of flood protection, and the Department of Housing Development plays an important role of housing reconstruction during disaster recovery.

13.5.4 Capacity Building

While discussions on institutional arrangements to respond to all-hazards vary, the necessity of strengthening capacity of disaster management is clear. Staff at disaster and emergency management offices in the Cabinet Office and Cabinet Secretariat are transferred to and from other ministries and offices every few years and cannot develop their careers in disaster and emergency management. Accumulating institutional knowledge is difficult due to frequent staff transfers. A mechanism needs to be established, such as creating career paths and developing the capacities of the selected core staff at these ministries and offices. This can help ensure that disaster management knowledge and expertise are maintained and can be leveraged effectively in future emergencies.

The COVID-19 pandemic has highlighted the need to strengthen collaboration between natural disaster and epidemic management. Health sector staff, who are engaged in managing the pandemic, are specialists in public health but not emergency management. Resources, knowledge, and experience on natural disaster management should be used for managing epidemics (Ishiwatari 2020).

13.6 Conclusions

Establishing approaches to manage the cascading effects caused by disasters is crucial in promoting sustainable development under changing socioeconomic situations and climate. This study clarified the issues of adopting all-hazards approaches by examining changes in policies and institutions in Japan. The country has established no comprehensive legislation of an all-hazard approach for emergency management but manages them practically based on the segregated laws and institutions of managing natural disasters, nuclear accidents, influenza and epidemics, and armed conflicts. The Cabinet Secretariat is leading in responding to emergencies affecting human lives and assets through coordinating various governmental organizations and utilizing the professional knowledge and expertise accumulated by the line ministries of infrastructure, health, agriculture, science, and education, and industry government agencies. In addition, four offices in the government are separately mandated as specialized agencies managing these hazards.

Establishing mechanisms for staff development and career paths can strengthen capacity and accumulate institutional knowledge to manage the cascading effects caused by disasters. Collaboration between natural disaster and epidemic management is also essential in strengthening disaster management in Japan.

Arranging governmental institutions requires careful consideration to promote an all-hazard approach to disaster management. Transplanting the FEMA-type organization may not fit with the Japanese bureaucratic system.

Acknowledgements This study was partially supported by the Japan Society for the Promotion of Science KAKENHI, Grant Number 21H03680.

References

Alcántara-Ayala I, Burton I, Lavell A, Mansilla E, Maskrey A, Oliver-Smith A, Ramírez-Gómez F (2021) Root causes and policy dilemmas of the COVID-19 pandemic global disaster. I J Disaster Risk Red 52:101892

Alexander D (2018) A magnitude scale for cascading disasters. I J Disaster Risk Red 30:180–185

Association of Prefecture Governors (2019) Recommendation on mechanisms of disaster recovery and risk reduction following the recovery period. Association of Prefecture Governors.

Central Disaster Management Council (2005) Business continuity guidelines. Cabinet Office, Tokyo.

Chen HY, Das A, Ivanov D (2019) Building resilience and managing post-disruption supply chain recovery: lessons from the information and communication technology industry. Int J Infor Manag 49:330–342

Consultation group for Japanese disaster management (2017) Report on Japanese disaster management: recommendation on establishing disaster management agency. Union of Kansai Governments. https://www.kouiki-kansai.jp/koikirengo/jisijimu/bosai/1164.html

Council of Vice-ministers concerned with examining emergency management institutions of the government (2015) Emergency management institution of the government. Cabinet Office, Tokyo.

DeWit A, Shaw R, Djalante R (2020) An integrated approach to sustainable development, national resilience, and COVID-19 responses: the case of Japan. I J Disaster Risk Red 51:101808

Fire and disaster management Agency (FDMA) (2017) White paper on fire management. FDMA, Tokyo.

Galanakis CM, Rizou M, Aldawoud TM, Ucak I, Rowan NJ (2021) Innovations and technology disruptions in the food sector within the COVID-19 pandemic and post-lockdown era. Trends Food Sci Technol 110:193–200

Hata I (2022) Research on bureaucracy: Indestructible group of making Japan (in Japanese). Kodansha, Tokyo

Hoeppe P (2016) Trends in weather related disasters–consequences for insurers and society. Weather Clim Extrem 11:70–79

Ide A (2021) Tourism and ICT solutions in the COVID-19 era: a comparison between Japan and Sweden. Rev Socionetwork Strat 15(1):195–211

Inoue M, Arakida M, Paudyal YR, Razak KA, Tsao TC, Ghosh C, Ishiwatari M (2022) Building disaster resilience amid the COVID-19 pandemic: a transdisciplinary approach for decision making. J Disaster Res 17(1):144–151

Ishiwatari M (2021) Institutional coordination of disaster management: engaging national and local governments in Japan. Nat Hazards Rev 22(1):04020059

Ishiwatari M, Koike T, Hiroki K, Toda T, Katsube T (2020) Managing disasters amid COVID-19 pandemic: approaches of response to flood disasters. Prog Disaster Sci 6:100096

Ishiwatari M, Ranghieri F, Taniguchi K, Mimura S (2021) Learning from Megadisasters in Japan: sharing lessons with the world. J Disaster Res 16(6):942–946

Ishiwatari M (2020) Multi-hazard approach to compound/cascading disasters: putting experience in developing disaster risk reduction to use in pandemics. JIIA research report. https://www.jiia.or.jp/en/column/2020/12/research-reports-global-issues03.html

Jamieson J, Yamashita N, Epstein DA, Chen Y (2021) Deciding if and how to use a COVID-19 contact tracing app: influences of social factors on individual use in Japan. Proc ACM Hum-Comp Interact 5(CSCW2):1–30

Japan Business Federation, KEIDANREN (2021) Toward establishing socio-economy resilient against emergencies: based on experience of COVID-19 pandemic (in Japanese). Japan Business Federation, Tokyo

Japan Science Council of Japan (2020) How can communities become less vulnerable despite intensifying disasters (in Japanese). Japan Science Council of Japan, Tokyo

Kato S, Yamaguchi Y (2019) Expected direction of Japanese public health emergency preparedness based on a review of similar procedures in the United States. J Natl Inst Public Health 68(3):230–249

Kishimoto A (2015) Integrated risk governance by "all hazard approach" (in Japanese). ITE Technical Report 39(15):11–16

Kjekstad O, Highland, L (2009) Economic and social impacts of landslides. In: Landslides–disaster risk reduction. Springer, Berlin, Heidelberg, pp 573–587

Leo B (2012) Can Japan respond better to its next large disaster? The Asia-Pacific J Japan Focus 10(21)

Machidori S (2015) Large-scale disasters and decision-making structure: considering from the role of FEMA (in Japanese), In: Consideration of wide area support during disasters. Hyogo Earthquake Memorial 21st Century Research Institute, Kobe, pp 113–126

Ministry of Environment (2020) Impact evaluation report of climate change, synthesis report (in Japanese). Ministry of Environment, Tokyo

Miyagi Prefecture (2014) BCP guideline for companies in Miyagi (in Japanese). Miyagi Prefecture, Sendai

Mori H (2019) Follow-up of recovery agency and continuing organization. (in Japanese). Legislat Surv 419:73–83

Mutai S, Leo B (2004) Making capacity of disaster management higher (in Japanese). Gyosei, Tokyo

Nagata S (2022) Recent trends of all-hazard crisis management system in Europe (in Japanese). J Soc Saf Sci 12:3–20

Nikkei (2021) Statement of Tokyo epidemics: not to repeat a crisis (in Japanese). Nikkei

OECD (2018) Towards an all-hazards approach to emergency preparedness and response, Boulogne-Billancourt. Nuclear Energy Agency Organisation for Economic Co-Operation and Development, Boulogne-Billancourt

Okamura M (2012) Japanese local disaster management capacity comparing with systems in the United Sate of America and the United Kingdom (in Japanese). Reference 7–25

Pescaroli G, Alexander DE (2016) Critical infrastructure, panarchies, and the vulnerability paths of cascading disasters. Nat Hazards 82(1):175–192

Pescaroli G, Alexander DE (2015) A definition of cascading disasters and cascading effects: going beyond the "toppling dominos" metaphor. Planet@Risk Glob Forum Davos 3(1):58–67

Ranghieri F, Ishiwatari M (2014) Learning from megadisasters: lessons from the Great East Japan. World Bank, Washington DC

Reich MR (2020) Pandemic governance in Japan and the United States: the control-tower metaphor. Health System Ref 6(1):e1829314

Sasaguchi, Y (2014) The cabinet in a parliamentary system: perspectives on the form of the Governance System in Japan. Legislat Surv 348

Tashiro A, Shaw R (2020) COVID-19 pandemic response in Japan: what is behind the initial flattening of the curve? Sustainability 12(13):5250

Yu Z, Razzaq A, Rehman A, Shah A, Jameel K, Mor RS (2022) Disruption in global supply chain and socio-economic shocks: a lesson from COVID-19 for sustainable production and consumption. Oper Manag Res 15(1):233–248

Chapter 14
All-Hazards Approach at the Local Level in Japan

Miwa Abe, Takako Izumi, Kumiko Fujita, and Rajib Shaw

Abstract What kinds of responses to various hazards are seen at the community level? In Japan, not only are responses to hazards complicated and intertwined with various ministries and agencies, but also, especially in the case of natural disasters, there is a widespread belief that "it's a disaster, so it can't be helped" even in the case of compound disasters, and the reality is that hazards are not being discussed in combination. In this chapter, we use the July 2020 compound disaster in Ashikita Town, Kumamoto Prefecture, as a case study to summarize what hazard approach issues were highlighted in the affected area. This disaster was a compound disaster caused by a COVID-19 pandemic, a heavy rain disaster, and a factory explosion accident (one of the Natech cases), and was a case that required a hazard response that had never been experienced in Japan before. Based on interviews and a literature review, this study summarized the situation of response to the combined disaster from the standpoints of the government, residents, and companies, and examined what the issues are at the community level when considering the all-hazard approach. As a result, it became clear that there is currently no system in place to enable residents to recognize the risks of compound disasters that exist in the community, and that the risks that each department has control over are not integrated because the departments that deal with the risks are functioning in different ways due to the effects of stove-piped government. This is the reason why the current situation does not integrate the risks that each department has in its control. In attempting an all-hazards approach, risk communication among the government, companies, and residents is essential, and it is necessary to share the hazards in the community. Therefore, it is necessary to have a system to share hazards from a bird's eye view of the entire region.

M. Abe (✉)
Tokai University, Kumamoto, Japan
e-mail: abe.miwa.k@tokai.ac.jp

T. Izumi
Tohoku University, Sendai, Japan

K. Fujita
International Consortium On Landslide, Kyoto, Japan

R. Shaw
Graduate School of Media and Governance, Keio University, Tokyo, Japan

T. Izumi et al. (eds.), *All-Hazards Approach*, Disaster Risk Reduction,
https://doi.org/10.1007/978-981-97-1860-3_14

Keywords The heavy rain in July 2020 · Natech · Risk communication · Compound disaster

14.1 Introduction

The Sendai Framework adopted at the 2015 UN World Conference on Disaster Risk Reduction emphasizes that its scope should apply not only to natural disasters but also to all disaster risks posed by anthropogenic and biological hazards and risks, and that national disaster reduction efforts should be conducted from an all-hazards perspective (Center Asian Disaster Reduction 2015).

The concept of the all-hazards approach as "emergency management" is said to be a method of responding to all disasters with the same principles of action, regardless of the type, cause, or scale of the disaster, whether man-made or natural, and is used in the same sense in this chapter (OECD 2018; Nagare 2017; Fukuda 2020). In modern society, the actors that practice crisis management are diverse, including "public assistance" led by the government and local governments, "mutual assistance" by companies, NPOs, and volunteer groups, "mutual assistance" by neighborhood communities and families, and "self-help" to protect oneself. Fukuda (2020) noted that although crisis management practices are required at all these levels, the actors driving crisis management in modern society are still government and business. While it is true that government agencies are the driving force behind hazard response, if effective risk mitigation is to be considered, it may be necessary to consider how the community side, including mutual aid and self-help, can recognize hazards that exist in the community and take action to address them.

The purpose of this study is to focus on actual compound disasters that have occurred and to examine, based on case studies, what challenges exist not only from the perspective of governments and administrative agencies as policy makers, but also from the perspective of private companies and local communities when considering the need for an all-hazards approach as a countermeasure for such compound disasters that may occur in the future.

The case study was a heavy rain disaster in Kumamoto Prefecture in July 2020. This heavy rain disaster was a compound disaster involving a COVID-19 pandemic and a factory explosion, which required administrative staff to respond to a hazard they had never experienced before. For the residents of the area, this disaster must have been their first experience with flooding and explosions during an infectious disease pandemic. The study was conducted through interviews with the town office of Ashikita Town, Kumamoto Prefecture, where the flooding damage and factory explosion accident caused by the heavy rain were confirmed, and through a literature review.

14.2 Definition of Terms

14.2.1 All Hazard Approach

All hazards are considered "A process, phenomenon or human activity that may cause loss of life, injury or other health impacts, property damage, social and economic disruption or environmental degradation (UNDRR 2020)". Regarding the all-hazards approach, in which the same principles of action are applied to all types, causes, and magnitudes of disasters, including man-made and natural hazards, OECD (2018), in considering nuclear emergency management systems, recognizes the importance of the all-hazards approach in preparedness and response. OECD also noted that an important step in achieving an all-hazards approach to emergency management is to consider the experience of emergency management of hazards arising from different sectors. This chapter defines the all-hazards approach as a method based on the OECD concept of responding to all hazards with the same principles of action, regardless of the type, cause, or scale of the hazard, such as man-made or natural disasters.

14.2.2 Compound Disaster

In disaster response, overlapping disasters in the same area are likely to cause severe damage or a shortage of human and material resources due to overlapping disaster responses. Such a situation is a "compound disaster," which can be classified into two types: a "simultaneous disaster," in which a series of disasters cause severe damage, and a "compound disaster," in which multiple disasters are simultaneously responded to, resulting in a shortage of resources. In particular, "simultaneous response compound disasters" are caused by a lack of human and material resources due to the simultaneous response to different disasters, resulting in a delayed response (Nakabayashi 2020). Another definition of a compound disaster is "a disaster in which another natural disaster (subsequent disaster) occurs in the process of recovery from a natural disaster (preceding disaster), causing more damage than the simple sum of the damage caused by each disaster alone (Itagaki et al. 2015)." The definition of a compound disaster in this chapter is a continuous and simultaneous disaster in which a subsequent disaster occurs in the course of a preceding disaster, causing more damage.

14.2.3 Natech (Natural Hazard Triggered Technological Disasters)

Natech is an industrial disaster triggered by a natural disaster, as opposed to a normal industrial accident caused by machine failure or human error. In Natech cases, it is said that a multidisciplinary approach is needed to consider risk assessment and risk management. We are now at the stage where we should consider risk management systems at the national level risk communication at the corporate and community levels, and the associated preparedness to respond to crises. According to Nakabayashi's classification, this can be categorized as a "simultaneous response compound disaster".

In the EU, studies have been initiated on how natural disasters can induce technological disasters and how the simultaneous occurrence of these disasters can be a tremendous risk for countries and regions that are not prepared for such disasters (Vetere Arellanoet al. 2004). In Japan, research on Natech has begun, but most papers examine the Natech case in the Great East Japan Earthquake of 2011 (Yu and Hokugo 2015; Aliperti et al. 2017). In addition to the direct effects of the industrial accident, Natech's effects on the consumption of agricultural and marine products due to the accident at the Fukushima Daiichi nuclear power plant have also been confirmed (Vetere Arellano et al. 2004). In compound disasters such as Natech, there is also concern about the impact on these primary industries.

In the next section, we will take a closer look at why the all-hazards concept is necessary in disaster response, based on a case study of a compound disaster that occurred in Japan, and examine the current situation where the concept of the all-hazards approach is not widespread, despite the fact that we experience natural disaster damage every year.

14.3 Example of a Compound Disaster in Kumamoto Prefecture in 2020

The July 2020 flooding disaster in the southern region of Kumamoto Prefecture occurred in the midst of an infectious disease outbreak, and overlapping hazards were identified, including a factory explosion caused by flooding. This section looks in detail at the damage caused by each of the hazards that formed the compound disaster and the agencies that responded to the hazards.

14.3.1 Flood

Ashikita Town, Kumamoto Prefecture, the target of this study, is located in the southern region of Kumamoto Prefecture (Fig. 14.1). In 2020, there were 102,959

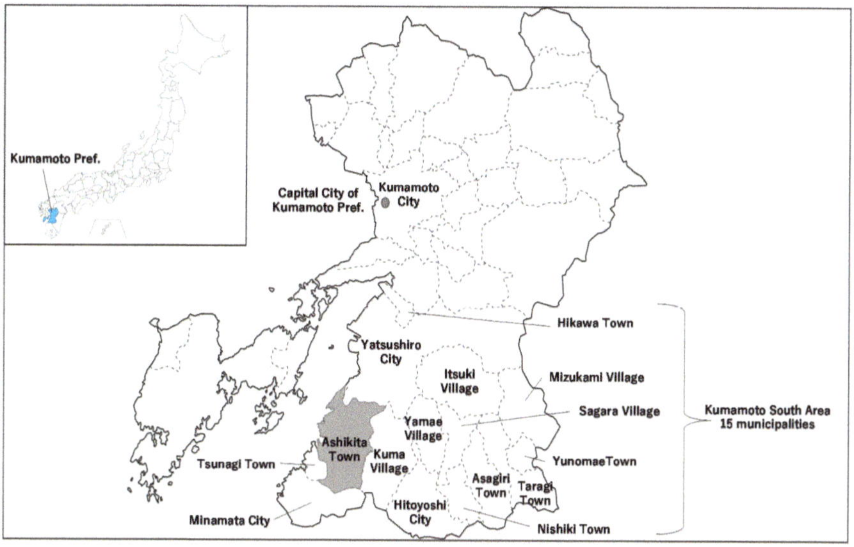

Fig. 14.1 Areas covered in this study (prepared by the author)

households and 259,133 people living in the southern part of the prefecture (Statistics Bureau of Japan 2020). The Kuma River, the largest first-class river[1] in the prefecture, flows through this area, which has a basin area of 1,880 km2 and a population of approximately 140,000. Flooding occurred in the Kuma River basin in 1965, 2005, and 2006 (Kyushu Regional Development Bureau 2023).

From July 3 to 8, 2020, the rainy season front remained stagnant, passing near Kyushu (southernmost of the four main islands of Japan) and extending into eastern Japan. This resulted in heavy rainfall in western and eastern Japan, especially in Kyushu, which experienced record-breaking rainfall from July 4 to July 7. According to information from the Fire and Disaster Management Agency at 13:00 h on July 4, an evacuation order (emergency) had been issued for 199,044 people and 90,222 households in four cities, three towns, and four villages in Kumamoto Prefecture, and an advisory for 71,530 people and 33,024 households in four cities, one town, and one village. On the morning of July 5, the next day, an evacuation order (emergency) at evacuation level 4 was issued to 204,567 people and 92,703 households in three cities, four towns, and four villages in the prefecture (Cabinet Office 2021).

As of January 7, 2021, 84 people were dead, 2 were missing, and 77 were injured, for a total of 163 casualties nationwide, and 1,621 houses were completely destroyed and 4,504 half destroyed. Among the victims, Kumamoto Prefecture suffered the

[1] Under the River Act enacted in 1965, water systems of particular importance for land conservation or the national economy and designated by government ordinance are called “First-Class Water Systems.“ Of the rivers in First-Class Water Systems, those that must be managed under the River Act and designated (with limited sections) by the Minister of Land, Infrastructure, Transport and Tourism are called “First-Class Rivers”.

greatest fatalities, with 65 people killed in the prefecture. The damage in Kumamoto Prefecture is shown in Table 14.1. The Kuma River, which flows through the southern region of the prefecture, overflowed, causing much damage (Fig. 14.2).

14.3.2 Covid-19

This flooding was unique in that it was a torrential outbreak amid concerns about the spread of COVID-19. Unprecedented evacuation center management methods were required, such as alcohol disinfection and space to avoid "three-densely packed" evacuation centers. The government also provided push-type assistance, masks, disinfectants, and other infection-prevention supplies in addition to the usual daily necessities (Cabinet Office 2020).

As of July 14, after the flooding, a total of 49 people (including 3 deaths) were confirmed to be infected in Kumamoto Prefecture, and countermeasures against infectious diseases were in the midst of being explored. Subsequently, on July 29, a Self-Defense Forces member dispatched to the affected area was found infected. The difficulty of taking countermeasures against infectious diseases in disaster relief activities in the affected area and how to manage volunteer activities in the affected area became an issue for the affected area.

The operation of evacuation centers differed from previous disaster area activities in that the thermograph (Fig. 14.3) appeared at the entrances of evacuation centers, the reception desk was centralized in one location, partitions were set up at a distance, and volunteers' activities were restricted. Restrictions on the acceptance of disaster relief volunteers across prefectural borders resulted in the inability to gather volunteers in the affected areas. Hitoyoshi City, the hardest-hit area, and other affected areas in the Kuma River basin are adjacent to Kagoshima and Miyazaki prefectures. In fact, Hitoyoshi citizens live closer to Kagoshima City than to Kumamoto City. Even considering the use of airports, the nearest airport would be Kagoshima Airport, but in consideration of COVID-19, volunteers were recruited only from within the same prefecture. It can be said that it was a very difficult time to strike a balance between securing personnel for the restoration of the affected areas and countermeasures against infectious diseases.

14.3.3 Factory Explosion (Case of Natech)

This torrential rain disaster caused a steam explosion and fire at a plant in Ashikita-town, Kumamoto Prefecture, Japan, when rainwater entered the graphitization furnace as a result of flooding.

A fire (explosion) occurred in the graphitization furnace equipment building at TOKAI CARBON's Tanoura Plant at around 11:44 a.m. on Saturday, July 4, 2020. It was presumed to be a steam explosion and fire caused by rainwater entering the

Table 14.1 Damage in Kumamoto prefecture

Municipalities	Human victimization		Housing Damage			
	Deceased / unknown party	Casualty	Complete	Half	One part	Flooding above/ below floor level
Kumamoto city					1	6
Yatsushiro city	5	19	147	160	97	
Hitoyoshi city	20	19	900	1,443	286	473
Arao city				133	105	
Minamata city				11	108	
Tamana city				8	18	
Yamaga city	2	1		8	19	8
Kikuchi city			1	2	2	
Uki city					1	48
Amakusa city		4		24	232	184
Koshi city				1		
Nankan town				39	8	
Nagasu town						2
Nagomi town				1	27	15
Minamioguni town				9	29	
Oguni town			3	29	20	40
Ubuyama village					1	1
Nishihara village					1	
Minamiaso village						2
Kosa town						5
Yamato town					1	
Ashikita town	12		72	910	559	
Tsunagi town	3		4	12	89	
Nishiki town				64	75	
Taragi town			1	8	15	50
Yunomae town					41	1
Mizukami town				1	4	6
Sagara village			18	90	75	
Itsuki village			1			6
Yamae village			11	14	20	
Kuma village	25	1	332	74	51	

(continued)

Table 14.1 (continued)

Municipalities	Human victimization		Housing Damage			
	Deceased / unknown party	Casualty	Complete	Half	One part	Flooding above/ below floor level
Asagiri town				51	55	43
TOTAL	67	44	1,490	3,092	1,940	890

Source Prepared by the author based on data from the cabinet office

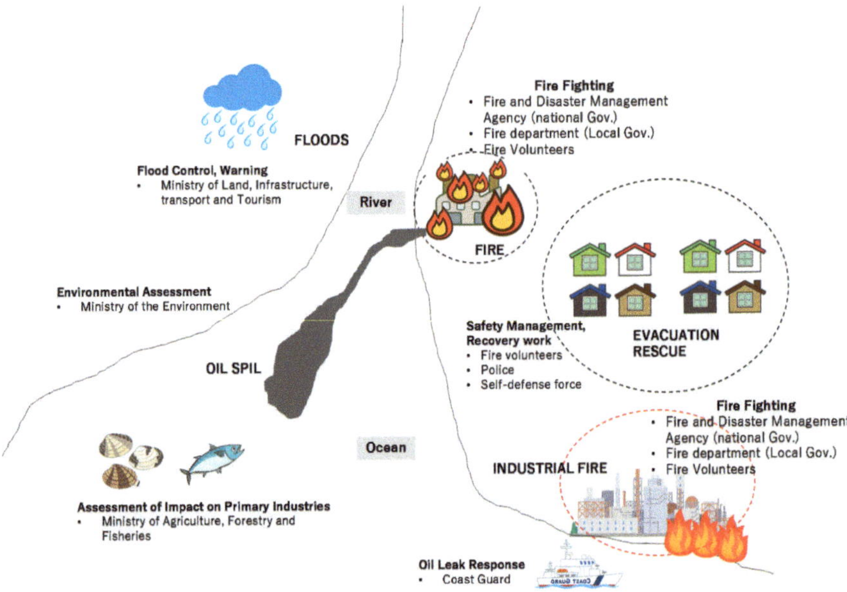

Fig. 14.2 Difference in water level between normal times and during flooding. (*Location* Kuma Village, Kumamoto pref.)

plant's graphitization furnace (2,000 °C) after the entire plant was flooded by heavy rains from the previous day. At 12:54 pm on July 4, 2020, the fire response was suspended due to the possibility of an explosion if water was released (Asahi Shimbun Digital 2020), and the fire finally subsided at around 6 am on July 5, the next day, and at 2:30 pm the fire was extinguished by the fire department (Minamata Ashikita Wide-area Administrative Union Fire Department). Fortunately, no human casualties were reported. However, it was announced that it would take time to restore the graphitization furnace that caught fire and that the restoration of production facilities other than the graphitization furnace, which had been shut down due to flooding in the plant, was scheduled for late August. While the impact on the supply chain was said to be limited, the impact on operations was expected to last two months.

This plant had suffered a similar steam explosion due to flooding in 1982, and a telephone interview with the Ashikita Town Office officials revealed that they were

Fig. 14.3 COVID-19 measures newly seen in evacuation center. (photo by the author)

aware of the possibility of another explosion since a similar accident had occurred in the past. As similar river overflows had caused explosions in the past, we asked the local government what they were doing to deal with hazardous material facilities and the residents living in the vicinity of such facilities. The local fire department would be the one to know the type and quantity of hazardous materials. There is no place for risk communication between local residents and companies, and the position is that safety management of hazardous material facilities is solely the responsibility of each company. In a hearing with Ashikita Town General Affairs Division officials, the actual sharing of information on the fire between the administration and the fire department was limited to the time and circumstances of the fire, the time the fire was extinguished, and the number of firefighters involved, and was handled in the same manner as fires during normal times.

14.4 Challenges for Each Stakeholder in a Compound Disaster

14.4.1 Challenges in Risk Sharing Between Communities and Businesses

Natech risks are often overlooked in regional and national risk assessments (Girgin et al. 2019). It is noted that in a compound disaster such as Natech, the economic and social conditions of the industrial facility and the surrounding area may make it more difficult for emergency responders to deal with the situation than in a typical chemical accident, and that the general public needs to be involved in a comprehensive discussion (Steinberg et al. 2008). However, the term "Natech" itself has not been generalized in Japan, and has not penetrated even the government agencies that respond to disasters (Luoet al. 2020). The difficulty in dealing with Natech is that the general public, the private sector, and the government are not familiar with the risks associated with Natech in Japan. Hazard maps are one tool that can be used to consider disaster response at the community level, but while they may show predictions of new areas of water damage, they do not show the risks posed by hazardous material facilities.

For example, a review of the hazard map for Ashikita Town, as in other cases, identifies the location of hazardous material facilities, but does not confirm any mention of the Natech risk posed by these facilities. The legend of the hazard map for flooding in Ashikita Town shows only Special Warning Area (mudslide), Warning Area (mudslide), Special Warning Area (steep slope failure), Warning Area (steep slope failure), Warning Area (landslide), mudslide danger zone, steep slope danger area, landslide danger area, and mudslide danger stream, but no risk related to hazardous material facilities. No risks related to hazardous material facilities are indicated. In addition to public facilities such as town halls, police stations, schools, banks, temples, and shrines, only gas stations are included in the legend as hazardous material facilities, although the depth of inundation during a flood is indicated from less than 0.5 m to more than 20 m. This is true not only for Ashikita Town but also for other local governments, none of which reflect the location of hazardous materials facilities or other chemical plants in the area as a risk in their hazard maps.

Although the plant was not confirmed to have caused any damage to residents in either the 1982 or 2020 floods, hazard maps confirm that this plant is located close to a train station and a residential area. The nearest train station is about 200 m away, and a residential area extends on the other side of the station. Despite the risk of explosions triggered by flooding, the factory is located in an area of 0.5 m to less than 3.0 m, suggesting the possibility of flooding, but this is not only a perception of Natech risk for the government, the company, and the residents, but it is also not linked to a response to this risk. The current situation is that it is not only perceived as a Natech risk by the government, private company, and residents, but also not linked to a response to the risk.

In the case of the flooding disaster, everyone understands that the plant explosion occurred as a result of the heavy rain disaster, but most people feel that the responsibility for the disaster is "unavoidable because it was a natural disaster." As for the risk response to a factory explosion, as evidenced by the perception that the response to a hazardous materials facility is not handled by the government but by specialists such as fire departments, the responsibility for risk management has not been clarified, as it is recognized that the location of risk management is a different department from that for natural disasters, even though it is a risk to which neighborhood residents may be exposed. It is also clear from the fact that hazard maps do not reflect the risk of hazardous material facilities. For both the government and residents, the "risk from natural disasters" and the "risk from industrial accidents such as factory explosions" are treated as two different things, and there has been insufficient effort to raise awareness that they are hazards that can occur simultaneously, or to establish sufficient opportunities for risk communication among the government, residents, and companies. Therefore, despite the experience of the Natech incident in 1982, neither the government nor the local residents share the risks.

14.4.2 Structural Problems in the Government's Response to Disasters

In this chapter, we have introduced the cases of compound disasters such as COVID-19, heavy rain disasters, and factory explosions (industrial accidents). Why, above all, has the government not been able to handle these various risks in an integrated manner? Let us consider this question from the perspective of disaster management organizations and the legal system in Japan. In Japan, where natural disasters are common, disaster response organizations and legal systems have developed in various ways. For example, Table 14.2 shows the laws of various organizations related to disaster response.

Table 14.2 Disaster response laws

Laws	Year
Disaster relief act (Act 118)	1947
Fire and disaster management organization act (Act 226)	1947
Japan coast guard act (Act 28)	1948
Fire services act (Act 186)	1948
Flood control act (Act 193)	1949
Police act (Act 162)	1954
Self-defense forces act (Act 165)	1954

Source Prepared by the author based on data from the Cabinet Office

The 2011 Great East Japan Earthquake caused damage to hazardous materials facilities and specified disaster prevention facilities in several petroleum compound and other special disaster prevention zones in the Tohoku and Kanto regions. In the Sendai area, a post-tsunami fire destroyed tanks and other storage facilities. In Chiba Prefecture, a fire broke out at a liquefied petroleum gas facility, exploding gas tanks and spilling thousands of kiloliters of oil at several facilities (Fire and Disaster Management Agency 2011). Coordination is underway to fight the fires and respond to the oil spills, which required cooperation across governmental organizations. The organizations that respond to disasters are under different administrative jurisdictions: the Fire and Disaster Management Agency of the Ministry of Internal Affairs and Communications for firefighting, the Ministry of Land, Infrastructure, Transport and Tourism for river management, meteorological services, and the Japan Coast Guard, and the Nuclear Regulation Authority is an external agency of the Ministry of the Environment. The Ministry of Land, Infrastructure, Transport and Tourism's Nuclear Regulatory Commission is now outside the Ministry of the Environment, while the Industrial Safety and Security Division of the Nuclear Safety and Security Agency has been transferred to the Information Policy Bureau of the Ministry of Economy, Trade and Industry (METI). For example, in the case of an oil spill from a factory that caught fire as a result of a flood-related industrial accident, the organizations in charge of responding to the fire, those in charge of dealing with the oil spill, and those involved in the industries affected by the oil spill would all have different responsibilities, so comprehensive disaster prevention measures and contingency plans would be necessary (Fig. 14.4). The response will also need to be considered from the national level to the local government level, and the private sector will need to actively participate in local disaster management.

14.4.3 Disaster Response and Business Continuity Challenges for Companies

The explosion and fire at TOKAI CARBON's Tanoura Plant, which was triggered by the heavy rain disaster, had caused similar steam explosions in the past due to flooding, and the community was aware that the plant might explode again this time.

Tokai Carbon's Tanoura Plant, where the explosion occurred, produces isotropic graphite materials (graphite materials for semiconductors, solar cells, and various industries). 120 employees and 45 subcontractors were working at the time of the disaster (Tokai Carbon 2020). It was the main plant for graphite materials, with an annual production capacity of 7,000 to 8,000 tons. Fortunately, there were no human casualties from the fire, and looking at the impact of the fire on the plant, the impact on the supply chain was limited, but the complete restoration of the graphitization furnace where the fire occurred "will take time in the order of years," and the company has fallen into a situation where there is no prospect of restoration. For private companies that have experienced accidents, it is extremely important to

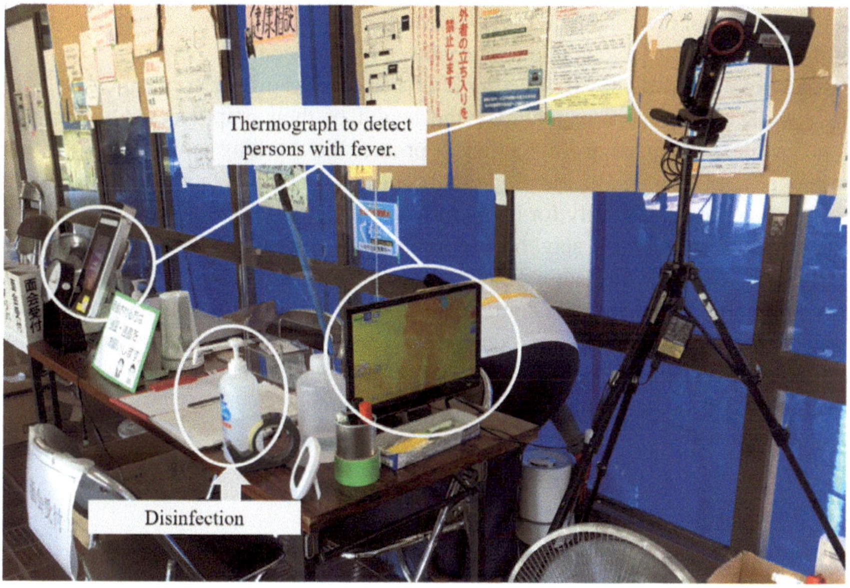

Fig. 14.4 Examples of agencies that would act in the event of an oil spill due to a factory fire caused by flooding (Created by the author)

formulate a BCP in order to continue their operations. In doing so, it will be necessary to plan not only for natural disasters, but also for various hazards, including infectious diseases, as we experienced with COVID-19.

In the case of the recent flooding, it was not simply a matter of planning for the risk of flooding at the plant. If the plant floods, there is a risk of an explosion, and they should have already experienced this in the past. Even though the damage was minimal, repeating a minor accident is nothing short of continuing to carry the risk.

There is even a case in Japan where a factory was closed as a result of a similar Natech matter as this factory. Most recently, a torrential rain disaster in Soja, Okayama Prefecture, in July 2018 caused a plant explosion caused by flooded water entering a high-temperature aluminum smelting furnace. The company that caused the Natech incident was an aluminum recycler established in Soja City, Okayama Prefecture, in September 1980. 5 people were rushed to the hospital and about 20 were seriously injured as a result of the accident. The explosion also scattered hot molten aluminum, damaging the roofs and walls of houses and causing fires in some areas, which burned down four nearby houses and warehouses. In September 2021, three years after the flooding, the plant that caused the explosion filed for bankruptcy protection at the Okayama District Court Kurashiki Branch and received a decision to commence proceedings. According to the trustee in bankruptcy, total liabilities amounted to approximately 5.4 billion yen, of which approximately 3 billion yen would be for obligations related to compensation for the victims (Sanyo Newspaper 2021). Thus, the Natech disaster is not only a case of recovery from a natural disaster but also a case of debt and bankruptcy due to the burden of compensation for victims of the accident and accidental damage caused by the accident.

The experience of COVID-19 after 2020 has also led private companies to review their work environment to maintain the health of their employees. On the other hand, while companies are obligated to protect the working environment of their employees, no laws can be identified that clearly define the relationship with neighboring residents. Especially for companies with hazardous material facilities, daily risk communication with neighboring residents is essential, and when considering BCP, the location of the factory and its relationship with the surrounding community should be considered as a continuity factor, in addition to business partners and other related organizations and consumers. Opportunities for risk communication with the local community on a daily basis would help reduce risk in the event that the company becomes a party to an industrial accident.

14.5 Conclusion

Focusing on the compound disasters that occurred in 2020, we have examined, based on case studies, what challenges exist not only from the perspective of governments and administrative agencies as policy makers, but also from the perspective of private companies and local communities when considering the need for an all-hazards approach to deal with such compound disasters that may occur in the future. The issues identified from the case studies in this chapter are as follows.

- Even though cases of industrial accidents triggered by natural disasters have already been observed in Japan, they are not understood by people such as governments, companies, and local residents. Therefore, preparedness for compound disasters is not discussed or prepared.
- As a result, government responses to industrial accidents that occur in conjunction with natural disasters and infectious diseases such as COVID-19 are recognized as separate risks even if they occur at the same time as natural disasters, and are handled by different departments with different jurisdictions, and there is no department that comprehensively handles the risks.
- It is imperative that local residents not only focus on natural disasters, but first become aware of the need to identify hazardous material facilities in the community, understand the risks, and work with businesses on disaster prevention measures.
- The government will need to consider a system that allows such private companies and local residents to share a common perception of risk, as well as the cross-jurisdictional cooperation necessary for disaster response as well as disaster recovery.
- The all-hazards approach is an important concept not only in policy making but also in considering business continuity in the private sector. Private companies should not limit themselves to their own disaster prevention measures, such as

developing a BCP, but also actively create opportunities for risk communication, such as conducting disaster drills in cooperation with the local community, so that they can better understand in advance the friction with the community in the event that they become a party to a compound disaster.

References

Aliperti G, Cruz AM, Rizzi F, Frey M (2017) Risk and Uncertainty: A Side-Effect of a Natech Event on Residents in Japan, Annual Report of Disaster Prevention Research Institute, Kyoto University B, 60(B), 335–342

Asahi Shimbun Digital (2020) 12:54 pm. https://www.asahi.com/articles/ASN744475N74TLVB019.html. Accessed 19 Dec 2022

Cabinet Office (2020) Damage caused by the torrential rainfall in July. https://www.bousai.go.jp/updates/r2_07ooame/pdf/r20703_ooame_33.pdf. Accessed on 20 May 2023

Cabinet Office (2021) Damage caused by the heavy rainfall in July 2020. https://www.bousai.go.jp/updates/r2_07ooame/pdf/r20703_ooame_40.pdf. Accessed 14 Dec 2022

Center Asian Disaster Reduction (2015) Sendai framework for disaster risk reduction 2015–2030

Fire and Disaster Management Agency (2011) Ministry of internal affairs and communications, 2011 white paper on fire fighting. https://www.fdma.go.jp/publication/hakusho/h23/cat-2/2/555.html. Accessed 03 Jan 2022

Fukuda M (2020) Philosophy of the all-hazards approach in crisis management studies (KIKIKANNRIGAKU NI OKERU All-Hazard Approach NO RINEN, in Japanese). Risk Management Studies 4:4–17

Girgin S, Necci A, Krausmann E (2019) Dealing with cascading multi-hazard risks in national risk assessment: The case of Natech accidents. Int Disaster Risk Reduct 35:1–13

Itagaki O, Hattori A, Fukuhara N (2015) How should multiple disaster be considered in measures against natural disasters? Research Trends and Results, Annual Report of NILIM, p 45

Kyushu Regional Development Bureau (2023) Ministry of land, infrastructure, transport and tourism, status of river improvement kuma river. http://www.qsr.mlit.go.jp/n-kawa/r_kuma/situation.html. Accessed on 25 June 2023

Luo X, Cruz AM, Tzioutzios D (2020) Extracting natech reports form large databases: Development of a semi-intelligent natech identification framework. Int I Disaster Risk Sci 11:735–750

Nagare H (2017) Disaster-resistant city planning for expected huge earthquakes (YOSOUSARERU KYODAIJISHINNNI SONAETA TAISAINO MACHIDUKURI, in Japanese). Environ Res Control 39:2–8

Nakabayashi K (2020) Combined disaster preparedness under multiple disasters and COVID-19. Sci Fire Disaster Prev (SAIGAITAHATSU TO SHINNGATAKORONA MANENKA NO FUKUGOUSAIGAITAISAKU, Published by SHOUBOU TO BOUSAI NO KAGAKU; in Japanese) 142:42–47

OECD (2018) Towards an all-hazards approach to emergency preparedness and response, Boulogne-Billancourt. Nuclear Energy Agency Organisation for Economic Co-Operation and Development, Boulogne-Billancourt

Sanyo Newspaper (2021) Explosion caused by torrential rain bankruptcy of south aluminum company proceedings to be initiated (*GOUU DE BAKUHATSU SOUJYA NO RUMIGAISHA HASAN TETSUDUKI KAISHI KETTEI*, in Japanese)

Statistics Bureau of Japan (2020) Census, basic aggregation of population; e-Stat. https://www.stat.go.jp/data/kokusei/2020/kekka.html. Accessed on 25 June 2023

Steinberg LJ, Sengul H, Cruz AM (2008) Natech risk and management: an assessment of the state of the art. Nat Hazards 46:143–152

Tokai Carbon (2020) Fire at our Tanoura plant (Second Report). "Japan Rubber Weekly" newspaper. https://gomuhouchi.com/materials/29233/. Accessed 07 June 2024

UNDRR (2020) Hazard definition & classification reviews, technical report

Vetere Arellano AL, Cruz AM, Nordvik J-P, Pisano F (eds) (2004) Analysis of Natech (natural hazard triggering technological disasters) disaster management. In: NEDIES Workshop proceedings. European Commission Joint Research Centre

Yu J, Hokugo A (2015) Household evacuation behavior during a Natech accident, vol 7. Memories of the Graduate School of Engineering and System Informatics Kobe University, pp 1–7

Chapter 15
Information Sharing to Build Localization into All-Hazards Approach to Humanitarian Assistance and Disaster Relief

Leigh Sholler, Alan Aoki, Michelle Ibanez, and Alberto Morales Jr.

Abstract Local and civil society organizations are the bearers of societal memory and retain a permanent presence in communities. In case of a natural or man-made disaster, local stakeholders will not only be the first (and enduring) responders but may also provide external stakeholders with avenues to deliver relief to communities affected by disasters. Those external stakeholders need to have built accountability, trust, and transparent mandates in order to successfully construct a relationship with local stakeholders. It is this relationship that will allow national and global networks mobilized for preparedness and emergency response to do the most good. Information sharing among local, national, and international stakeholders is the foundation of that relationship, and the success or failure of building that relationship will determine the type of engagement external players have with local stakeholders before, during, and after an emergency. Information sharing is thus a keystone to building resilience. Although trust deficits and lack of accountability have disrupted historic disaster responses, more recent disruptions have come from externalities, as seen during the COVID-19 pandemic, during which some communities in the Asia–Pacific dealt with natural disasters mostly in isolation and largely in the absence of direct international support due to public health restrictions. There is a breadth of literature on civil-military coordination at the international level, but it is limited at the national and, more importantly, local level. It is our intention to suggest ways to shore-up information sharing shortfalls to ensure that localized preparedness and response benefit from broader community expertise and networks. As disaster management institutions and practices shift toward an "all-hazards" model of preparedness—building functions and practices that address the risks and impacts of various hazards rather than focusing wholly on the facets of specific hazards—information sharing that ensures the centrality of local actors will be ever more important.

L. Sholler · A. Aoki (✉) · M. Ibanez · A. Morales Jr.
Applied Research and Information Sharing (ARIS) Branch, Center for Excellence in Disaster Management and Humanitarian Assistance (CFE-DM), Hawaii, USA
e-mail: Alan.Aoki.ctr@cfe-dmha.org

L. Sholler
e-mail: leigh.sholler@cfe-dmha.org

T. Izumi et al. (eds.), *All-Hazards Approach*, Disaster Risk Reduction,
https://doi.org/10.1007/978-981-97-1860-3_15

Keywords Localization · Trust/accountability · Information sharing · Resilience · Disaster risk reduction · Civil-military coordination · All-hazards approach · Information technology · COVID-19

15.1 Introduction

Although thirty years have elapsed since localization became an explicit driving concept in disaster management, information sharing is among the aspects of localization that remain unelaborated. Difficulties remain in ensuring that the international, regional, and national levels of preparedness and response share information and resources with local actors to ensure that communities have what they need to bolster their own resilience in the face of multiple, evolving hazards. Information sharing is a critical component of preparedness and response as without information communities are prevented from fully participating in preparedness and response activities.

The need for inclusion and empowerment of national and local actors in humanitarian assistance and disaster response has been recognized by larger international humanitarian organizations and donors who have committed to international strategic initiatives for increased localization. The Grand Bargain and the Charter for Change are the most prominent of these initiatives. Since localization was made a formal priority in the 2016 World Humanitarian Summit's "Grand Bargain," there has been increasing consensus at policy and normative levels, underscored by the Coronavirus Disease 2019 (COVID-19) pandemic, that local leadership should be supported. Yet, change on the ground remains slow.

There are varying definitions of localization, ranging from a narrow meaning of making funding directly accessible to local organizations, to building the capacity of local actors and institutions, enhancing the use of local institutions as implementing partners, or advancing locally led humanitarian response and development (Cooley et al. 2021). This chapter focuses on locally led disaster responses in a permissive environment but also offers considerations for local- or community-led disaster risk reduction (DRR) with an all-hazards approach. While all responses are led by the Affected State, large-scale international assistance can supplant local mechanisms and sideline the meaningful participation of local and national actors in decision-making. The emphasis on locally led responses prioritizes local and national leadership in decision-making, as facilitated by actions to increase local capacity and resources. The most substantive and empowering approaches are toward locally led development that supports local institutions in a manner that promotes sustainability, reflects current ethical sensibilities, incorporates the voices of vulnerable groups, and prioritizes the perspectives and preferences of recipient governments, local civil society organizations, and host country professionals (Bonis-Charancle and Vielajus n.d.). The overall trend is to move away from large, international response mechanisms and as much as possible toward more local response capabilities, which are more responsive to and reflective of local needs and can, therefore, be mobilized to confront all of the various hazards to which a community is exposed.

This chapter explores information sharing at the international, national, and local levels with the goal of filling some gaps in the literature surrounding best practices for information sharing to ensure that localized preparedness and response benefit from broader community expertise and networks. Concurrently, the international community can benefit from learning from localized preparedness and response. These relationships with the local or affected community at the center must also be built into civil-military processes and policies that prepare potential external responders ahead of a humanitarian assistance/disaster relief (HA/DR) mission.

15.2 Literature Review

In 1994, at the First World Conference on Natural Disasters, held in Yokohama, Japan, attendees endorsed the Yokohama Strategy and Plan of Action for a Safer World, under which Affirmation 6 stated:

> Community involvement and their participation should be encouraged in order to gain greater insight into the individual and collective perception of development and risk, and to have a clear understanding of the cultural and organizational characteristics of each society as well as of its behavior and interactions with the physical and natural environment. This knowledge is of the utmost importance to determine those things which favor and hinder prevention and mitigation or encourage or limit the preservation of the environment for the development of future generations, and in order to find effective and efficient means to reduce the impact of disasters. (IDNDR (International Decade for Natural Disaster Reduction) 1994)

The signatories went on to stress in Principle 6 that:

> Preventive measures are most effective when they involve participation at all levels, from the local community through the national government to the regional and international level. (IDNDR (International Decade for Natural Disaster Reduction) 1994)

Participants in the Conference then laid out strategies to integrate local and traditional knowledge and expertise with scientific and technical knowledge to ensure the fullest buy-in and greatest potential to mitigate risk and reduce the impacts of natural disasters (IDNDR (International Decade for Natural Disaster Reduction) 1994).

By the time the 2004 Indian Ocean ("Boxing Day") Tsunami struck, academics, policymakers, and humanitarian organizations had begun to try to understand how affected communities could be supported before, during, and after disaster not only because of the role they physically play but also because of the role their knowledge can play. After-action reviews of the 2004 tsunami indicated that some adverse outcomes could have been avoided if national, regional, and global stakeholders had not only sought local information but had also integrated local actors (community leaders, civil society organizations, etc.) into planning and preparedness activities (Dekens 2007). Dekens (2007) faulted large, global institutions for failing to reflect upon the shortcomings of technical knowledge and top-down command during disaster events and suggested that decentralizing DRR and climate change adaptation (CCA) away from central government (and security) apparatuses and toward local entities will ensure that local, context-specific information is available and applicable rather than simply feeding a centralized planning process wherein all local knowledge gets generalized and diluted.

In the wake of the Indian Ocean Tsunami, the development of the Hyogo Framework for Action (HFA) sketched out a 10-year blueprint (2005–2015) for DRR that involved building institutions, identifying and assessing risk, enhancing early warning, building a "culture" of safety and resilience, and reducing underlying risk (Hyogo Framework for Action 2022). Adopted by 168 countries and endorsed by the United Nations (UN) General Assembly, HFA came up for reconsideration as 2015 drew near and stakeholders began to craft the "post-2015" DRR agenda. In its preparatory review for the 2015 Third World Conference on Disaster Risk Reduction, also slated for Japan, United Nations Office for Disaster Risk Reduction (UNISDR) consulted with countries and other stakeholders to gauge progress on the HFA priorities and to assess appetite for a follow-on or revamped DRR program (UNISDR (UN Office for Disaster Risk Reduction) 2013). Among the loudest conclusions of the pre-2015 consultation was that the post-2015 framework had to retain a focus on local action—i.e., well-resourced and well-supported local governments, community engagements, and community ownership of effort (UNISDR (UN Office for Disaster Risk Reduction) 2013). While these findings emphasized linkages of policy, law and regulations, participatory planning, and information sharing (UNISDR (UN Office for Disaster Risk Reduction) 2013), these stakeholder views appear to have been superseded by ideas that "localization" was related mostly to funding resources.

At its own endorsement by the UN General Assembly after the 2015 Third World Conference on DRR, the Sendai Framework for Disaster Risk Reduction 2015–2030, attempted to refocus DRR on state-centered action with consultation among and buy-in from local governments, the private sector, and others (UNDRR (UN Office for Disaster Risk Reduction) n.d.). However, none of the four Sendai Framework Priorities centered local actors—either governments or community organizations. While one of the guiding principles called for delivering "resources, incentives, and decision-making responsibilities *as appropriate* (emphasis added)" to local authorities and communities, no targets were incorporated to measure whether or how this principle was followed (UNISDR (UN Office for Disaster Risk Reduction) 2015). In something of a backlash against this apparent shortcoming, a significant shift in how to conceive of local control of disaster management, humanitarian action, and other related activity erupted in 2016.

The Grand Bargain, launched during the World Humanitarian Summit in Istanbul, Turkey, in May 2016, brought some of the world's largest donors and humanitarian organizations together in a commitment to get more resources into the hands of people who are at risk or in need as a way to improve the effectiveness and efficiency of humanitarian action (IASC (Inter-Agency Standing Committee) n.d.). While the Grand Bargain was primarily focused on delivering more robust and more reliable funding to local actors, it also suggested a way forward for informational elements of DRR to be changed into a more equitable system.

In 2021, the Grand Bargain signatories agreed that the process needed to evolve, and they endorsed the framework of the Grand Bargain 2.0 whose overall objective is:

> Better humanitarian outcomes for affected populations through enhanced efficiency, effectiveness, and greater accountability, in the spirit of Quid pro Quo as relevant to all. (IASC (Inter-Agency Standing Committee) n.d.)

Although one objective remained to ensure that more funding reached local players, a workstream of the Grand Bargain 2.0 is that greater support be provided for the "leadership, delivery, and capacity of local responders and the participation of affected communities in addressing humanitarian needs" (IASC (Inter-Agency Standing Committee) n.d.). This revival of the Grand Bargain responded to criticism that international donors and organizations were focused primarily on funding local implementation of internationally-planned projects, and it was deemed crucial that these major players plan for and encourage local leadership and participation at all stages of development, DRR, and HA/DR actions.

As early as 2017, local stakeholders in several Pacific Island countries (PIC) reported that they expected localization to bring greater effectiveness, efficiency, economy, and equity but were mostly concerned with timeliness that could only be achieved by recruiting and resourcing local staff (Ayobi et al. 2017). Indeed, rather than being most concerned with funding streams, local PIC stakeholders consistently expressed expectations that localization would be a process of evolution of local, national, and international systems and structures that would eventually ensure national and local actors have other stakeholders' respect and support to ensure the most effective humanitarian response (Ayobi et al. 2017). Carlson (2017) pointed out that every DRR framework from Yokohama through Sendai (2015–2030) as well as Pacific community strategies has incorporated a plank promoting local knowledge to inform policies and plans.

Among United Nations Office for Disaster Risk Reduction (UNDRR) projects aimed at bridging gaps in the information sphere between local and top-down humanitarian action, DRR, and CCA is the "Words into Action" series that promotes best practice in localization. In 2019, the UNDRR stated that community involvement is not only about tapping local knowledge and resources, but it is also about understanding how communities make choices according to their opportunities and constraints; community organizations are on the ground during collective responses and use their own networks and support systems (UNDRR (UN Office for Disaster Risk Reduction) 2019). The most recent focus has been on including traditional knowledge to bolster the sustainability of DRR at local levels (UNDRR (UN Office for Disaster Risk Reduction) 2021) because:

(1) Locals can take ownership of DRR processes
(2) Impacted people are the best guides for relevant solutions
(3) Local/traditional knowledge takes local capacity into consideration
(4) External interventions and technologies that integrate local knowledge are more likely to be accepted and sustainable.

Information that will ensure efficient and effective DRR—covering safety, health, or even relocation—must be both accessible and comprehensible to the targeted communities in terms of language, cultural framework, and technology or medium (UNDRR (UN Office for Disaster Risk Reduction) 2019).

UNDRR (UN Office for Disaster Risk Reduction) (2019) enumerated the qualities necessary in a system of sharing of information, resources, and human skill that would underpin successful DRR and CCA:

(1) Redundancy–spare capacity purposely created to accommodate disruption and multiple ways to fulfill a particular need
(2) Robustness–accommodate failures and ensure that failure is predictable
(3) Flexibility–willingness and ability to change, evolve, and adapt in response to changing circumstances
(4) Reflectiveness–learning from the past and adjusting
(5) Resourcefulness–rapidly find alternative ways to achieve goals or meet needs
(6) Inclusivity–emphasize the need for broad consultation and participation
(7) Integration–as the process brings together systems, institutions, and actors, it can envision additional benefits as resources are shared and actors work together

Among those efforts that are beginning to prod global, regional, and national stakeholders into more inclusive action are surveys such as that conducted by the Global Network of Civil Society Organisations for Disaster Reduction (GNDR) in 2019. The survey research, "Why are people still losing their lives and livelihoods to disasters?" found that, among other shortfalls, disaster-affected communities and communities trying to undertake DRR and CCA planning efforts are excluded even by their local governments (GNDR (Global Network of Civil Society Organisations for Disaster Reduction) 2020, Conclusion 2). Fewer than one-third of community members surveyed by GNDR (Global Network of Civil Society Organisations for Disaster Reduction) (2020) said they were included in evaluations of DRR projects and that their exclusion meant that they were unaware of resources available before, during, and after disasters. Moreover, responses from local community actors and local governments consistently show differing priorities (GNDR (Global Network of Civil Society Organisations for Disaster Reduction) 2020).

GNDR (Global Network of Civil Society Organisations for Disaster Reduction) (2020) cites the example of the Pacific Climate Change Science Program in Timor-Leste in which communities, local partners, international organizations, and governments shared both scientific and local knowledge for DRR and CCA in a way that ensured that community members had access to information about local impacts of climate change and could incorporate scientific and local knowledge into improving the sustainability of their livelihoods. The refrain remains the same: without information, communities are prevented from fully participating in DRR processes and activities. National governments and other non-local stakeholders must systematically engage citizens and take into account that disability, language, and choice of communication medium can hinder even the best-intended project.

There were early indications that the COVID-19 pandemic may have stimulated systemic changes within the global humanitarian, DRR, and CCA communities that other efforts—e.g., the Grand Bargain—had not (Barbelet et al. 2020). With international donors and actors limited in where they could travel for the better part of two years, local action became conspicuously the norm globally rather than the exception (Barbelet et al. 2020). Funding and supplies moved globally without a robust human accompaniment while capacity building and information sharing occurred remotely in an example of how things could be in a post-COVID world if stakeholders remained committed to building trusting partnerships upon which they will rely in both future crisis and during preparedness and risk reduction projects.

Alas, the 2022 State of the Humanitarian System (SOHS) report revealed that humanitarian stakeholders' pandemic experiences were not entirely rosy. Remote, international management and decision-making remained the rule while local implementing partners were expected to overcome mobility and communication challenges, accepting all the risks of failure without any opportunity for driving strategy, planning, or resources (ALNAP 2022). The upshot was a palpable loss of trust between local organizations and their target communities as well as between local implementers and international or regional organizations. The SOHS particularly pointed out that local actors were unable to combat pandemic-related and general misinformation because of limits imposed on both their use of international donor funding and their movement among target communities (ALNAP 2022). While the SOHS suggests that bolstering two-way communication between humanitarians and affected communities is the most basic way out of this morass of mistrust:

> Failures to consistently 'close the feedback loop' by providing a response to aid recipients is creating a trust deficit between aid recipients and agencies that threatens the uptake of consultation and feedback mechanisms (ALNAP 2022).

What is more, the SOHS found that the pandemic may have exacerbated rather than alleviated the seemingly inherent power imbalance in the relationship between humanitarians and the people they serve, who may be both local implementing partners as well as disaster-affected people (ALNAP 2022). This imbalance stems from the fact that humanitarians participate in the relationship by choice while necessity or circumstance presses affected communities into it.

"Localizing" humanitarian action within this context would seem absolutely necessary, but the process itself remains fraught as pandemic-pressured international donors and organizations appear to only accept local actors who model Western-dominated values and frameworks because that nominally shows that locals can be trusted with data and funding. If international agencies "localize" without devolving power and resources to local communities, as the SOHS indicates was the dominant schema during the COVID-19 pandemic (ALNAP 2022), few changes to decision-making and planning will emerge. If, instead, other models of localization can be adopted and implemented, then all-hazards preparedness and HA/DR missions can be more effective.

Going forward, stakeholders are beginning to examine models for localization to incorporate improved accountability, power-sharing, and information flows within the nascent "green humanitarian" movement. Although the overall goal of such efforts is to ensure that the humanitarian system does not exacerbate the climate change crisis, a clear consequence of such a shift will be to uproot colonial power dynamics that observers note remain embedded within the humanitarian system and that marginalize indigenous and local knowledge (HAG, Glow and PIANGO 2022). In international humanitarianism, as in the broader climate change discourse, the under-representation of indigenous voices at global conferences contributes to "global North"-led approaches that not only may fail to save lives and reduce suffering but may also contribute to environmental degradation, less resilience, and additional risks and vulnerabilities (HAG, Glow and PIANGO 2022). The "green humanitarian" movement foresees a future wherein more seats at the table are held by locals, indigenous communities, or representatives of the "global South" and, therefore, information shared among stakeholders is of higher quality and is more useful for elaborating strategies and programs to confront existing and future, climate change-influenced hazards and to ensure that humanitarian action does not exacerbate risks and vulnerabilities, including to environmental collapse.

15.3 Methods

This research uses a qualitative approach to gather stakeholder insight into this complex challenge. Phenomenon-driven research using a grounded theory approach was used. The process involves two key tasks: (1) a review of literature related to the history and evolution of localization and the challenges surrounding information sharing in this context; and (2) data collection through semi-structured interviews with practitioners from across the HA/DR community with representation from the varying levels of preparedness and response to include international, regional, national, and local levels. Conducting interviews provides the opportunity to collect empirical material from a wide range of individuals working in the HA/DR community.

Pursuant to Schwarz and Stensaker (2014), phenomenon-driven research focuses on recognizing, securing, documenting, and hypothesizing a phenomenon of interest in order to "facilitate knowledge creation and advancement." It is significant for theory building (Von Krogh et al. 2012). Early phases of phenomenon-based research are best supported by qualitative research, as there is no theory that can be tested (Edmondson and McManus 2007).

Purposive sampling was used to identify interviewees. Purposive sampling is the process of selecting a sample by taking a subject that is not based on the level or area but is based on the specific purpose (Arikunto 2010). The pool of candidates for our research is the community of HA/DR stakeholders to include the private sector, government, international, national, and local non-governmental organizations (NGO), academia, and international regional centers. Thus, we have selected individuals and organizations based on their status as prime stakeholders in implementing commonly accepted localization initiatives. Interviewees are at the core of localization as they are responsible for implementing the strategy in their day-to-day activities and decisions.

From a potential pool of twenty-four humanitarian practitioners invited to participate in the interview process, a total of ten individuals from seven organizations did sit for in-depth interviews. The interviewees were mostly selected through professional contact networks and by snowball sampling (Miles and Huberman 1994). Interviewees were asked to suggest other individuals with expertise on our research topic either from their or another organization. The researchers interviewed one or two experts from each organization in order to enable triangulation of the results within the same organization (cf. Alvesson 2011). Many of the interviewees had major responsibilities related to information sharing, with titles such as Regional Coordinator, Humanitarian Affairs Officer, or similar (40%). Some focused on information sharing and localization with job titles such as Disaster Risk Resilience Advisor, Information Sharing Chief, or Regional Information Management Officer (30%). The study also included three interviewees (30%), titled Senior Advisor and Executive Director who were considered subject matter experts rather than working for a specific organization and had extensive, general experience working in the sector.

The interviewees were highly experienced, having worked in the humanitarian sector between four and thirty years with a mean of eighteen years. While the greater part of the interviewees currently work at senior leadership levels, many of them have extensive fieldwork experience in disaster response, mitigation, information sharing, and capacity building, with most still deploying in response to humanitarian emergencies. Their combined fieldwork experience covers a large geographical area. The areas most experienced by interviewees were Southeast Asia, the Middle East, Europe, and the Pacific Islands. A majority of the interviewees work or have worked in the humanitarian sector. To secure the anonymity of the interviewees, we cannot provide further details; all interviewees' views are marked by an "I"—for interviewee—followed by a numeral, and their pronouns default to a third person plural (they/them). Table 15.1 offers some detail on the interviewees' organizations, positions, and years of service.

Full disclosure was made to the interviewees with regard to the reason for the study, confidentiality, and the opportunity for the interviewees to aid enhanced practices and policies concerning information sharing in order to build and enhance localization. The researchers conducted the interviews via Microsoft Teams in teams of at least two researchers. The questions posed to each interviewee were identical and addressed in the same order. The series of questions covered topics such as definitions of localization, impacts of COVID-19 on operations and information sharing, information systems and technology uses and methods, policies and procedures, coordination mechanisms, and organizational philosophy as it relates to the integration of local/traditional/indigenous knowledge. The researchers utilized Excel to code each interviewee's statements in order to identify themes and commonalities in the answers. Findings from the qualitative portion of the survey were ascertained by using the triangulation method. Pursuant to Creswell (2011), this approach allows qualitative inquirers to triangulate among different data sources to enhance the accuracy of the study. In the second stage of data analysis, narratives and emergent themes were mapped determining key points or trends and how interviewees' institutions aimed

Table 15.1 Interviewee organization, position, and experience

	Organization level (international, national, regional, local)	Organization purpose	Interviewee position	Interviewee years of service at their current organization	Interviewee years of experience in the humanitarian field
I1	National	Academic	Associate professor	20	30
I2	Regional	Regional HA/DR coordination	Assistant director for monitoring and analysis	5	10
I3	National	Academic	Director, office of external affairs and advisor	9	4
I4	Local	NGO	Senior advisor	25	25
I5	International	Civil-military coordination	Humanitarian affairs officer	17	20
I6	Regional	NGO	Regional manager, pacific	8	22
I7	International	Civil-military coordination	Regional information management officer	15	21
I8	International	Civil-military coordination	Humanitarian affairs officer (Civ-Mil coordination & access)	3	12
I9	International	Civil-military coordination	Information management officer	8	9
I10	International	Supporting humanitarian response	Advisor/trainer	6 months	22

to support information sharing. The results of the data analysis are documented in Section IV of this research study.

15.4 Findings

Interviewees are steeped in the drive to "localize" DRR and HA/DR activities and have experienced its successes and failures. When asked to focus on the information aspects of localization, they were keen to discuss the ways in which data and information flow—or do not flow—within humanitarian operations as well as in pre- and post-disaster planning and review periods. There are optimists, like I2 who expect local and national capacity to strengthen in the coming years to the point that their regional organization sees requests for relief resources decline because national resources and expertise will be sufficient; rather their organization's role will be one of coordinating or serving as a knowledge hub. There are pessimists, like I1 who sees localization as a "fancy" international concept, and I4 who suggested that the "Grand Bargain" was made "about the global South" rather than "with" local partners. I4 further stated,

> The whole design of the info infrastructure was built for global North to fortify accountability according to them to measure relevance and transparency according to them; imagine a traffic cone, the North looks through the small part at us in magnification; that is not equal; what we say is inadequate and they view the global South as incompetent.

And there are realists, like I8 who cautioned against localizing "for localization's sake" and insisted that international organizations must find locals "that share our principles and standards." Given the variety of ways that "localization" is described and the fact that climate change-influenced hazards are causing and will continue to cause more intense disasters, further debate about resources, expertise, standards, and tools will remain fraught. Nonetheless, the desire to ensure that humanitarian actors at all levels of the community have access to information and the tools to create, analyze, and use it appears to be a point of agreement. As an example, the COVID-19 pandemic was a significant event that shone a light on how the humanitarian community had—or had not—localized operations. The sudden, intense focus on virtual meetings, remote support, and locally led humanitarian action wreaked havoc in some organizations while others thrived. The whole world—not just humanitarian actors—experienced a new world of information access and technology, and those experiences have the potential to change how information is shared and used in future humanitarian operations.

What is Localization

Generally, practitioners interviewed agreed that "localization" means that local leadership and resources are used to address problems within a community. However, there are different definitions of localization, and they vary from organization to organization and even among individuals within an organization. Along with this broad variation, there does appear to be a divide between how "local" actors might define localization and how regional or international organization staff define it. Moreover, "local" actors may view it both as an international concept that seeks to implement outside ideas (standards) within a local context and as facilitating the strengthening of actors who have not historically had a voice. I10, who has many years of international agency and organization experience but now works in a private-sector organization supporting disaster relief, said that localization is "really the way ahead for the humanitarian space; making maximum use of the people, leveraging their in-country culture and how they operate to make sure operational input meets national and people's needs."

Some international organization respondents cautioned that a local solution does not always make it an ideal or even acceptable solution. I5, who works in an international organization says,

> Localization should not mean everything local is appropriate … I don't think localization should just mean that everything local is perfect, but I think it means genuine dialogue; that needs to happen -- respect for hosts.

Despite the diverse views of what localization is and means, all respondents advocated for the promotion of localization as an effective means of response during emergencies. I1, an academic in a PIC, stated,

> Localization is critical; we must ensure it is bottom-up rather than led by people trying to "train" people who already have their own way; that's why they have survived.

A key element in understanding the standpoint of some local actors is the history of suppression of local voices, particularly under colonial systems that devalued "local" actors and knowledge. At the same time, international organization respondents acknowledge this historical gap and indicated that they are aware of the discrepancies and consciously work towards diminishing this. I3, speaking about Bangladesh, said,

> It is struggling to get onto the stage, elbow its way up and say 'we do know what we are talking about; it is our land.'

And it is not just an issue of international actors looking down on local ones. I3 also pointed out that some local/national academics must be sold on value of local knowledge; I8 underscored that issue by highlighting that a select one or two privileged individuals who are "local" are invited over and over again to international conferences even though they may not represent a cross-section of the people of their country as they generally provide a privileged view from the top 1% of society. The national dynamics that drive this selection will be difficult for non-national and non-local organizations to push back against. Moreover, international organizations appear aware of the issue although they likely will continue to acquiesce to the selections of national governments when it comes to choosing representatives.

An issue that several interviewees broached was the sense that localization should not be considered the panacea or even a stand-alone solution for improving response, particularly when international humanitarian standards and laws directly clash with local practice. While recognizing that international actors cannot disregard local culture unless "we intend to fail," I8 also insisted that "there are times when the local culture or norm or practice is against our values" and brought up examples of female genital mutilation and child marriage. Additionally, for other areas of humanitarian assistance such as public health, interviewees brought up that local solutions to health matters sometimes do not mesh well with international health practitioner standards, i.e., would "western" medicine be more effective in treating a health issue or would a local method be sufficient? While useful to discuss, this question goes beyond the scope of this paper, as it moves away from the information sharing aspect of localization and delves more into the socio-cultural and moral/ethical issues of disaster relief. Nonetheless, it is worth noting that potential pitfalls of localization, both for the local and international practitioner, have not been thoroughly expounded upon.

On the issue of data and information sharing, national/local actors, I1, I4, and I6, expressed a sense that local communities and organizations feel mistrusted by international agencies. At the same time, international respondents expressed that the opposite was also likely true—that local populations needed to be convinced that they could trust the international community. The international humanitarian community appears keenly aware of the fact that there is a trust deficit between international responders and local responders and populations; they are, thus, highly cognizant of the fact that the international humanitarian community needs to work to decrease this divide. I9, who works in an international organization, brought up the importance of reciprocity in building trust:

> A lot of problems come from the fact that we ask for a lot of information and many times do not give anything back; our goal is to make sure if we get information what are our partners getting out of it… the goal is not just taking but to give back and listen to partners … make sure you give back.

In addition to attitudinal obstacles that may be interfering in implementation of localization strategies and plans, there is an institutional impediment in that many international/regional organizations have a mandate that dictates their work with specific levels of authority, most often the national government. Those with a specific mandate rely on other organizations to work with people or agencies beyond that mandate. I2, who works in a regional body, said their organization "supports at the national level NDMO [national disaster management organization], and it is their role to strengthen local" in what appears to be a reflection of most mandates as the preponderance of international and regional heft is focused on national-level authorities. I8 further elaborated in noting that information flows the other direction—from local to national or international agency—may be entirely absent.

> Very often, with limited access to affected people, we are not fully able to source the information from the affected population themselves; it is based on responders' analysis and understanding rather than what affected people are communicating.

The challenges inherent in not only accessing but also integrating and valuing local information were constantly apparent in the way that interviewees addressed any perceived or actual differences between "modern" scientific or technical information and local, indigenous, or traditional knowledge. In fact, some agencies—perhaps in a reflection of their societies—expect information to be vetted for applicability or accuracy at the highest levels before being shared. Nonetheless, all (ten out of ten) practitioners ascribed to the importance of both scientific/technical and local/indigenous information. But there were significant differences in the way they addressed the issue of using or valuing local information. The responses ranged from somewhat anodyne considerations of how different types of "data" could be integrated for analysis and processing by simply redefining how they are "counted" to a more substantive demand that, as I2 put it, indigenous knowledge be "reviewed scientifically." Two attitudes became evident: (1) local knowledge is valid because it is put in practice; and (2) local knowledge needs to be validated by scientific method.

The issue of local (encompassing traditional and indigenous) knowledge being integrated into the ways in which international, regional, and national humanitarian agencies operate was challenging to most interviewees, some of whom admitted to not having really considered the issue while others keep it at the forefront of their work. I10 noted that even amongst countries in a sub-region of the Indo-Pacific that may share socio-cultural ties, "people have a different thought process," which only adds to the complexities of this challenge. Above all, interviewees expressed the depth of the challenge they felt in ensuring that even if local knowledge was not "compatible" in form with standard practice, it could be complementary. Moreover, several respondents from within the information management (IM)/information technology (IT) professions highlighted that on-the-ground teams do not discount any seemingly accurate information—local or scientific—if it will save lives.

To overcome resource, policy, or attitudinal obstacles to localization, specifically in the information sharing realm, respondents suggested various remedies, but the most specific was to ensure all stakeholders have equal access to interoperable information systems. Several interviewees expressed confidence that "speaking the same language" in terms of data parameters, appearance, and access to platforms will level the playing field, and I9, an IM professional with an international organization, said, "tools depend on context, but we try to be flexible to whatever is most popular in [the space where we are working]." Of the three IM and IT professionals among the interviewee group, all underscored that the IM teams and working groups are very flexible and adapt to whatever the disaster preparedness or response requires, including using whatever platforms a local population is using (e.g., WhatsApp).

As will be discussed later in these findings, one of the major recent ways to assess what "localization" is and how it might work in the information sharing realm was the COVID-19 pandemic, during which response activities seemed to shift to national-/local-level control and, therefore, international and regional organizations relied on local and national individuals, organizations, and agencies for information on an emergency. I2 highlighted that, in disasters during the pandemic, their organization could not send out the quick response teams it would normally deploy; rather, "rapid needs and damage assessments were done locally… we saw that without external human resources in local areas, they could do the job."

What are the benefits and risks of sharing information

The reality of disaster response is that no organization works alone. Those interviewees who are part of an organization that deploys personnel to the field or to support emergency operations centers during a disaster response (all except I1 and I3) said, in various ways, that integrating other stakeholders is *"what we do."*

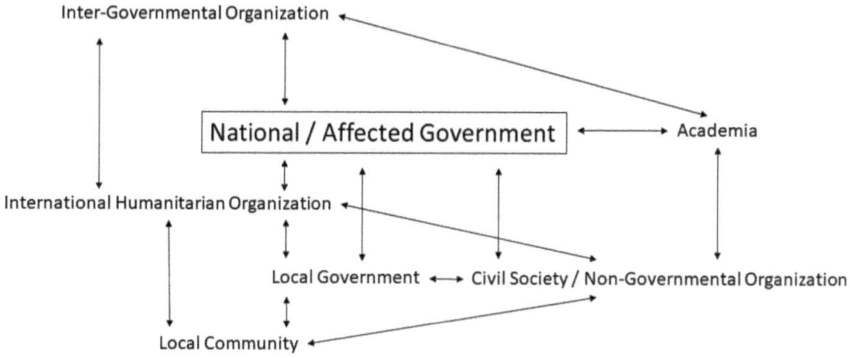

Fig. 15.1 Information flows among all stakeholders with national governments in central role during emergencies

However, mandates and legal restrictions can stand in the way. I2, I5, I8, and I10 in particular pointed out that:

> When we deploy or respond, it depends on the framework that is laid out by the affected government; that dictates how you engage… if the government platform is not inclusive, we are not inclusive (I5).

Some organizations do not have that mandate and rely on others that do, as I2 said, "our mandate is to work with national governments." I2's organization supports at the national level, and it expects national authorities to strengthen local response. This intermediary role of national authorities was strongly cited across interviewees, not usually as an obstacle to information sharing but as a necessary stakeholder in that national authorities retain legal oversight of disaster responses in most cases. Figure 15.1 illustrates the central role of national [affected] governments in participating and facilitating information flows among other stakeholders before, during, and after disasters.

This acceptance of legal mandates and stakeholder integration during disaster responses did not extend, in some interviewees' minds, to the realms of disaster preparedness and other non-emergency interactions; as I4 highlighted, locally-based civil society groups have minimal contact with top-level donors or international agencies outside emergencies. Thus, there appears to be a gap in relationships that contributes to a lack of trust when, during an emergency, responding groups need to share information. This gap in relationships could be filled in or at least reduced during non-emergency times.

All interviewees agreed that sharing most types of information at all times improves coordination, effectiveness, and efficiency. While I3 saw information sharing as a means to better society writ large, all nine of the remaining interviewees looked at more practical issues. I6 underscored that ensuring local actors have access to information produced by other stakeholders is increasing Pacific Islander communities' ability to integrate global approaches to disaster management into those they have already developed within their communities. It appears that information flows are a key way to break down geographical, linguistic, or socio-economic barriers to building resilience. Indeed, I2 explicitly revealed the desire not to limit this information sharing to the disaster management community but to bring academe/researchers together with practitioners. Practically, I9 pointed out that if information is not shared and shared from bottom-up and top-down then a disaster response will not be coordinated. Thus, by promoting accountability and transparency in collecting and disseminating information, organizations can get the "right information, sufficient information, timely information" and make better decisions (I7).

But placing just a wholesale value on information and ensuring its availability to stakeholders in a response is not entirely without challenges. I4 pointed out that the ability of humanitarian response practitioners and affected communities to deliver data and information from the ground via mobile devices has actually made it more difficult to reconcile vast amounts of data that sometimes run contrary to an official version of events.

> It seems the volume of information is not the problem; we have inflation of information; to discern meaningful information is the challenge. We are swimming in information, but we don't know the truth (I4).

This view was echoed by I10, who advocated thinking more critically about what information is needed, "what is essential information, what is desirable, and then, what is not needed." Moreover, I8 revealed the reluctance—or perhaps legal inability—of some international organizations to reach far down to the local level to gather and share information and said, "very often, we probably would reach more people if we rely on other actors to understand local dynamics." This reluctance may run the other direction as well; I1 and I10 suggested that for some NGOs, there may be restrictions on information they can share freely because it is their "bread and butter" (I1) or could impact their funding "and they get protective" (I10). At the same time, I8 underscored that international organizations also restrict information sharing on the basis of "sensitivity" of information to keep from exposing "individuals' identifiable information" to risk if it leaks into the public sphere.

Across the board, interviewees from international or regional organizations emphasized that they work with other organizations who have an on-the-ground presence and that this allows them to produce and disseminate information products. All but two respondents highlighted that on-the-ground teams should not discount any seemingly accurate information—local or scientific—because:

> When there is an opportunity to identify population needs… any information from the local context... is important to get into our [IM] tool (I7).

So, while data and information may be flowing relatively freely, how that information is assessed, processed, and publicized runs into lingering issues of "ownership." I4 again underscored that the global North continues to dictate the terms of "meaning" and "truth" while the global South must "justify what we say." Building trust among disaster management and humanitarian stakeholders remains a key to ensuring that data contextualization become de-colonialized and more equitable.

How do you share information

The practical aspects of sharing information and ensuring it reaches all stakeholders cannot be forgotten. Interviewees built on two separate issues on this score: policy and platform. On the one hand, an organization or individual may have formal or informal rules or guidelines about what type of information is shared and with whom it is shared, as well as with how to integrate information shared by others into organizational activity. At the same time, the technical aspects of information flows—digital, analog, face-to-face—can be an obstacle to sharing information if they are not thought out and managed ahead-of-time.

On the issue of policies or guidelines related to information sharing, some organizations have strong policies and rules for collecting and disseminating information, including the use of well-established tools, while others do it in an ad hoc manner. Three respondents pointed to formal and specific institutional projects ranging from the ASEAN Agreement on Disaster Management and Emergency Response (AADMER) work program and Inter-Agency Standing Committee (IASC) guidelines to internal guidelines and checklists as governing IM. Indeed, I2 underscored the "lengthy consultation with stakeholders" that went into regional programs for information gaps, challenges, and opportunities. At the other end of the spectrum, some organizations handle information flows in an entirely organic or ad hoc manner or based on the demands of their partners or host agencies. I4 said,

> We do the best we can… to be honest, we have not gotten that sophisticated; we don't have an SOP, and I6 underscored that if deployees' host agency says no sharing, we don't share, if they say please share, we share widely.

This is also true of national governments, who, as the affected state and default lead of response efforts, often dictate the level of information that is shared or can be shared. As I7 put it:

> In terms of IM, we take the government/local government lead; they are part of the people; so, we do take into account anything they provide.

In some ways, strictly adhering to the limits national governments may impose on information sharing may pose additional obstacles and challenges to response. A lack of timely information, or not enough information due to restrictions, can hinder relief operations and assistance. The results can include relief efforts not being directed to where they are most needed, duplication of efforts, delayed response, or a general lack of coordination, all of which could, in turn, lead to additional loss of lives and the prolonging of suffering of the affected populace. It is therefore essential for responders to have access to and exchange information during disaster relief operations in an efficient manner. Broadly, interviewees who work in or with international organizations tended to reference pre-determined guidelines while interviewees working at a local level work with information flows that reflect paths built before an emergency. In fact, all respondents recognized a need to have established relationships before a disaster response.

On this issue of building relationships before an emergency demands humanitarian action, respondents all agreed with the idea that relationships built before an emergency will encourage information sharing. I2, I6, and I10 emphasized that face-to-face contact before or during a crisis facilitates communication, with I2 saying that person-to-person relationships meant that "informal communications are easy and quick even before formal sharing is instituted" during a disaster response. I10 reiterated this, highlighting that even during the pandemic-era limits are on social interaction, "if you don't socialize and eat together, it doesn't work as well." At the same time, I6 cautioned that developing people-to-people ties required work to include a variety of people because using the same "selected one or two privileged individuals" may not represent a good cross-section of the group or community with whom information is being shared.

The personal and professional relationships built during pre-disaster workshops or networking events were particularly important to I1 whereas I2 saw less formal virtual discussion platforms as providing practitioners areas to work out ideas. In both cases, interviewees pointed to a mix of in-person and virtual means to communicate. Overall, interviewees all mentioned that their organizations and partners at the international, regional, national, and local levels use a variety of databases, software, and apps (email, the Microsoft Office suite, social media, etc.) to manage and share information. There does appear to be a difference between local use of digital platforms—especially WhatsApp—and international players' ability to rely on IT department-hosted tools and processes. This divide also appears to be reflected at the practical level where international-level staff and academics, I1, I2, I3, I5, I7, I8, I9, and I10, said meetings—in person and virtual—were absolutely unavoidable in humanitarian emergency contexts for decision-making to occur and, therefore, required IT department support. Indeed, I5 laughingly reiterated the "OCHA [the UN Office for the Coordination of Humanitarian Affairs] is a synonym for meetings" stereotype that points to how humanitarian responses are coordinated under formal structures.

Nonetheless, this digital sophistication among international or regional players is not absent at lower levels. IM and IT professionals among interviewees (I2, I7, and I9) emphasize that international IM teams are flexible and adapt to whatever the local population is using (usually WhatsApp) and that their respective agencies use common software and apps that can integrate or be shared with stakeholders outside their organizations. Indeed, all interviewees working in the international, regional, or academic spheres said that communication with external partners uses common tools—e.g., email, WhatsApp, social media, or even hard copy. Internal communications were more restricted to log-in protected platforms, such as Microsoft Teams and Sharepoint. I6 and I10 were outliers in pointing out that information in their organizations is ideally shared internally and externally through face-to-face encounters facilitated by the presence of staff in target communities. Figure 15.2 illustrates the types of tools most commonly used, according to our interviewees, during non-emergency (DRR, CCA, preparedness, after-action) activity periods versus those used during HA/DR responses.

While in declared emergencies that elicit a large-scale multi-national humanitarian response, national governments of affected states remain the nexus for most policy decisions, including regarding information sharing, respondents highlighted

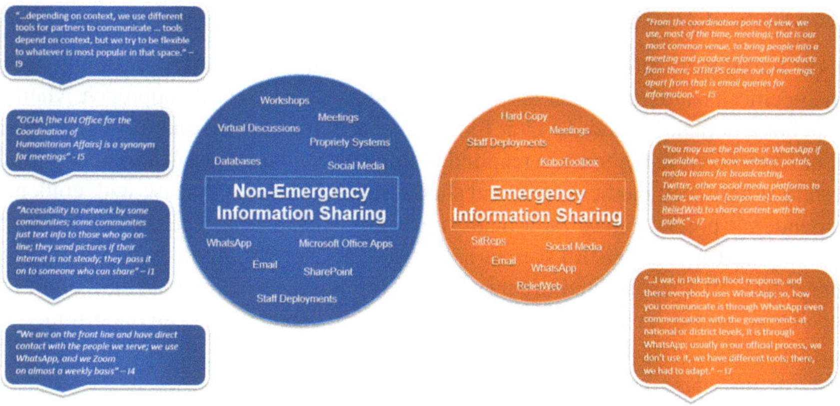

Fig. 15.2 Tools most commonly used to share information

that formal and informal structures link stakeholders at the international-national, national-academic, national-mass media, and national-local levels. These structures and the communication methods and preferences they feed differ by context, as highlighted by I7 who pointed to:

> Different tool[s], assessment tools, Kobo[Toolbox], people go to the field to talk to the community… you may use the phone or WhatsApp if available… we have websites, portals, media teams for broadcasting, Twitter, other social media platforms to share; we have [corporate] tools, ReliefWeb to share content with the public.

Three major challenges are still apparent despite practitioners' clear appreciation of the way digital (online) communication has expanded the boundaries of information sharing. One, I4 was the most forceful in pointing out a major problem with the plethora of venues and tools for data and information sharing without any investment on the front end to ensure that all stakeholders know what data to collect and how.

> My organization spent days and weeks collecting [data] only to be told the values were incorrect; the cost we paid for no payoff… it wastes time and energy (I4).

Two, the source, shape, and context of data and information can be overwhelming, or it may be a "difficult" piece of information that requires extra protection or contextualization. Different organizations collect, value, and utilize different types of information. International staff, especially I5, I7, I8, and I9, were emphatic that their organization[s] do not share "names, addresses, gender-based violence data" (I9). But it does seem that international staff respondents were more attuned to "sensitivity" and context (natural disasters vs. complex emergencies) while local actor respondents were more attuned to how to make information linguistically or culturally appropriate for local community audiences. Finally, the respondents were all interested in ensuring all stakeholders have equal access to interoperable systems in terms of language, technological infrastructure, and financing to collect and manage information.

Case Study—COVID-19 Pandemic

In the traditional worlds of disaster management and risk reduction, planning based on specific hazards has been shifting to a multi- or all-hazards approach wherein stakeholders plan for and develop functional capacities to respond to various types of hazards because the tools to respond to impacts can be similar—e.g., search-and-rescue, emergency medical care, and shelter management. A disease with epidemic or pandemic potential may traditionally have been addressed as a public health hazard that required specific health structures to plan and manage the response rather than integrating all agencies and organizations who may have functional skills into the plan for response. The COVID-19 pandemic was a challenge and an opportunity for further inroads in developing all-hazards approaches, including in how stakeholders manage and share data and information.

I10 illustrated how their organization "re-thought" processes when faced with the information sharing challenges of the pandemic, both on a technological level (e.g., how to effectively exchange information in areas of the Pacific with low bandwidth) and in terms of how and what type of information was necessary to share:

> [The pandemic] ...did make us ask how to share information; there was so much information; you had to set up platforms for information access; the problem with 'in the sky' platforms is ... you might not have the bandwidth to access it; so, you had to try to email it or share hard copies in outlying areas – if you are allowed in those areas; it made us think about what is essential information, what is desirable, and then, what is not needed; you become more targeted.

Across all interviewees, there was agreement that the key impact of the pandemic on HA/DR was a shift of response activities to the national and local levels. I2 was most blunt in saying "rapid needs and damage assessments were done locally," as even national governments could not send people, and "without external human resources in local areas, they could do the job." Even three years since the declaration of the pandemic, I3 observed that "you have to rely locally because internationals have not come back." All respondents noted that local resources became more heavily relied upon even as technology replaced or supplemented for the international absence.

Despite this broad agreement, interviewees' answers revealed that organizations felt disparate impacts on their operations depending largely on whether they deploy or not in response to disasters. I1, an academic, said a big change was that communities with whom they work "are not coming to us" for support in building resilience projects but rather their organization was reaching out to build partnerships. And I8 said,

> We at the regional office are called upon to deploy in emergencies; so, we were unable to physically deploy... the crux of the response had to be at the national level, and within certain countries there were movement restrictions.

There was no choice for most organizations but to "localize" in all respondents' views. What is more, however movement restrictions impacted their physical work, for all interviewees, there were changes in communication styles and methods based on both technology and new-found flexibility in time, formality, and hierarchies. I3, an academic, emphasized:

> Hundreds of thousands of lives are saved by technology and coordination; everybody gets involved; it seems to be phone, internet, on-the-ground, megaphones; WhatsApp is huge; everybody is on it.

All interviewees noted that pandemic era reliance on virtual (online) communication had benefits and drawbacks. Positive impacts ranged from lessened expenses (I1) to the opportunity for more people to participate in discussions surrounding humanitarian response (I7). At the same time, negative impacts ranged from exclusion of people without sufficient internet bandwidth (I1, I6) to increased pressure on practitioners to give more of their time because it was easier to connect (I1). I4 again pointed out that swift adoption of IT-based platforms meant more information but less time for analysis, and I6, I9, and I10 lamented that a lack of face-to-face interaction lessened the "human" side of humanitarian action. Moreover, I6 and I10 underscored that some parts of the world, particularly PICs, have both a cultural demand for face-to-face interaction and a technological shortfall that disrupts virtual communication; these elements have meant that people switched back quickly when movement restrictions ended.

Despite some positive outcomes—i.e., broadening the number of voices heard and reducing some of the costs of doing business—several interviewees reflected on the intersection of the COVID-19 pandemic and the localization strategy. Practically, I8 sees it as a "push for localization to be not just a buzzword" and to "source beyond the international humanitarian community to local non-humanitarians who have become humanitarians out of need." Indeed, I8 pointed out that while the UN system was "working out [how to use] Teams," NGOs already had local actors with whom they had long engaged, and they took on the work. But both I4 and I6 lamented the "accidental" (I4) or "last resort" (I6) nature of the move to rely on local actors for humanitarian action under pandemic restrictions. While people in the technology sector may have celebrated that their work over years of trying to connect people online had paid off, for actual control of resources, planning, and information, the pandemic was more of an externality that demonstrated the shortfalls of previous "localization" projects.

Concluding remarks

Our findings indicate broad acceptance of the need for local actors to have a voice in determining the shape and leadership of humanitarian responses and DRR activities that impact them. Moreover, all respondents agreed that all stakeholders in a humanitarian response require access to timely, accurate, and actionable information. However, there are diverse views among interviewees on how to make progress toward local leadership and how to ensure that stakeholders build relationships of trust within the humanitarian community to ensure that information flows from international to local and vice versa. Our questions to the interviewees sought to bring to light the elements that either promote or hinder information sharing among the various levels of humanitarian action—local, national, regional, and international. These elements include legal mandates, organizational policies, institutional structures, interpersonal relationships, IT tools, etc., and interviewees indicated the various formal and ad hoc ways that they manipulate tools to try to ensure humanitarian action is successful. I4 stated,

> We adapt new information a lot, maybe not with academic rigor, but we do the best we can; we adapt it as we see fit; to adapt our products, we use our website to share with our networks and present it in different platforms.

In the end, interviewees' comments began to sketch out several ways in which the humanitarian community can begin to practice more fluid information sharing before, during, and after a disaster and, therefore, how information sharing and the relationships that underpin it promote a functional approach to DRR that allows communities to build resilience to all types of hazards.

15.5 Recommendations

By speaking with practitioners and experts at various levels of the global humanitarian community—i.e., local, national, regional, and international—our study sought to tease out practices and considerations that promote information sharing during HA/DR operations and that can be generalized to an all-hazards approach to DRR. In particular, we sought to understand how to ensure that local stakeholders' knowledge, needs, and skills in information gathering, analysis, and dissemination could be promoted as part of a reinvigorated push to "localize" humanitarian action across all types of disasters. We purposefully did not focus on information sharing during complex emergencies due to potential legal limitations placed on information sharing within such a context.

Interviews with ten experts and practitioners showed that modern communications technology has had a significant impact on how information is shared before, during, and after a humanitarian response. However, interviewees' responses revealed a spectrum of positive, neutral, and negative impacts due to attitudes, policies, languages, and tools that differ among the various levels of assisting and affected entities. By ensuring that stakeholders take into consideration these aspects of communication in preparedness, implementation, and after-action review, they can make progress toward one of the goals of the post-2015 DRR agenda: well-resourced and well-supported local, community ownership of disaster management efforts (UNISDR (UN Office for Disaster Risk Reduction) 2015). Indeed, our interviewees' views seem to bear out Kelman's (2013) suggestion that as local communities begin to recognize that they are being relied upon to bear the costs of both losses and non-structural disaster management interventions, they will become louder about rejecting the centering of international and national agents in the DRR process. And local actors are not being shy about using modern communications technology to raise their voices.

All respondents agreed that information flows were crucial to successful humanitarian action regardless of the scale or scope of the operation. Indeed, there was minimal disagreement found across interviewees with regard to how and why information is shared with stakeholders, and the prevailing attitude was one of wanting to ensure good data and information get to the people who need it—both local communities dealing with response efforts and the international community supporting those efforts. Thus, we would like to propose recommendations centered around the previously mentioned four aspects—attitudes, policies, languages, and tools—that can, done well, promote effective information sharing, or done poorly, limit information flows and undermine the effectiveness of humanitarian action.

What are we recommending?

1. Build trust by ensuring two-way communication and including local actors as equal partners
2. Continuously advocate sharing information to break down hesitation
3. "Watch your language"
4. Use the information technology that is available to and used by the affected community

To address an overarching theme, based on respondents' general agreement that history continues to strongly influence the ways that "the global South" or less developed countries are treated by international entities rooted in "the global North," stakeholders in disaster management must be aware of historic, often post-colonial contexts and take care to not replicate colonial attitudes. The 2022 SOHS echoes our interviewees' feeling as two-thirds of SOHS informants felt that local actors' capacity needs and strengths are defined by international agencies without local input and that how international humanitarians treat affected populations can verge on the disrespectful amidst examples of international agencies using imagery of aid recipients for donor reporting and fundraising (ALNAP 2022). The global discussion on "decolonizing" all aspects of DRR, CCA, and HA/DR is on-going.

For the first of our recommendations, a major element needed to overcome gaps in information sharing and localization is to develop or foster trust between local communities and national and international stakeholders. Our interviewees were highly cognizant that trust was a requirement for information sharing at all levels: "We have to integrate trust building in everything we do" (I7). Trust-building of course, goes both ways, and will involve conscious effort by all stakeholders. International interviewees indicated sincere efforts towards ensuring that information sharing tools are aimed towards benefitting local communities. Tools in use by many international organizations include accountability mechanisms such as formal feedback processes, complaint boxes, or after-action interviews with both local implementers and target communities (ALNAP 2022). Local input can be elicited by institutionalizing two-way communication rather than keeping to one-way systems whereby international, regional, and national stakeholders tell local communities what DRR or HA/DR activity is occurring. Nonetheless, ensuring that target communities can react does not get to the heart of localization wherein local actors have a say in strategy, planning, and fund allocation. Progress toward this goal requires international stakeholders to, as I3 said, avoid devaluing "non-Western" knowledge and to, as I4, I7, and I9 said, ensure that data and information management platforms are available to and configured for use by local actors rather than for the benefit of the international donor community. International interviewees underscored efforts in this direction, but they also reiterated that they cannot always select the community members with whom they engage either because of their mandates or because social hierarchies within a country or community may stymie these efforts. It behooves stakeholders to try to build trust with non-elite community members as those actors selected by national disaster management agencies may be, as I8 indicated, privileged and not representative of the national public.

Second, all stakeholders in the international humanitarian community have policies or mandates that drive their actions and help define the partners with whom they engage. All interviewees reflected upon whether their organizations' policies were strict or loose on this front, but they all emphasized that the goal is for information to be as freely shared as possible. I2, in particular, pointed out a sticking point in that their organization's mandate explicitly connects it only to national government disaster management agencies and that they cannot reach beyond those partners without national government authorization and, in fact, rely wholly on national governments to pass down information to sub-national levels. I5, I8, and I10 also pointed out that their organizations hew to the framework laid out by an affected/host [national] government and that "if the government platform is not inclusive, we are not inclusive" (I5). Given the findings of a 2019 GNDR survey that affected communities and communities trying to undertake DRR and CCA planning efforts are excluded even by their own governments (GNDR, Conclusion 2), the whole community will confront legal and practical obstacles to sharing information. The best course of action in this case is to continuously advocate sharing information as part of a campaign to break down hesitation or inertia.

A complicating factor on this policy front is that it is not solely governments that will need encouragement to review their information sharing policies. I1 and I10 reflected that some organizations place limits on sharing their data and information because it is their "bread and butter" (I1) and gives them a means to earn funding or a seat at the table (I10). Disaster management practitioners and experts can maintain an advocacy campaign before, during, and after humanitarian assistance missions without tying it to specific situations or sensitive data. As part of campaigns to ensure all stakeholders understand the importance of sharing information to improve the efficacy of humanitarian action, the humanitarian community can take non-emergency periods to work toward common data standards and common definitions of "sensitive" data or information so that there is broader confidence in how partners will use information during an emergency.

Third, given the international context of HA/DR and DRR, there are multiple languages and multiple ways of communicating among stakeholders in various places and venues. Ensuring that information to be shared is appropriate, accessible, and actionable means taking into consideration vernacular, jargon, medium, and context. In other words, "watch your language." As I1 underscored, even the term "localization" can be considered a "fancy" international concept that local actors may interpret as outsiders trying to "train" locals who already live in their own space and have survived and, thus, do not need external expertise. I5 and I8 agreed that their international organizations probably need to have deeper conversations regarding how they talk to local communities and that they probably would reach more people by relying on actors who understand local dynamics. Key actions to be taken when managing information to be shared among various humanitarian stakeholders are: (1) ensure information is not crowded with international agency jargon; and (2) include information that is applicable to the local context. Of course, what is simply stated here will take international, regional, and national actors time to implement, but it requires intentional planning and execution to craft messages that the intended audience receives, understands, and can act upon.

Finally, with attitudes in mind, policies in place, and language adapted, there remains the issue of information systems, the tools of communication. There are, even in the world's most developed communities, financial and technical capacity limits on data and information systems. As illustrated by the Kingdom of Tonga's inability to communicate with the rest of the world after the Hunga Tonga-Hunga Ha 'apai volcanic eruption and tsunami of early 2022 destroyed submarine communication cables, these limits can be more conspicuous in less developed countries and in communities experiencing a disaster event. I1 and I8 pointed out some of the issues and the ways that people work around them with I1 indicating that while modern virtual communications can bring down the cost of doing business internationally, bandwidth in some remote communities is inadequate for even basic internet access. I8 echoed this by underscoring that many Pacific Island communities are easily cut off even from the 3G networks that connect them to the world and that, as a result, they continue to emphasize face-to-face communication regardless of whether a given agency or office has good communication infrastructure. I1, I8, and I10 report that people without internet access often pass information verbally (including by telephone or text message) to people who do have internet access in order to ensure that their data and information are included in national, regional, and international fora. However, in an emergency, this work-around can be too time-consuming. Our three interviewees who are IM or IT systems professionals underscore that, from their perspectives, humanitarian information management teams and working groups are very flexible and adapt to whatever the disaster preparedness or response activity requires, including using whatever platforms a local population is using, very often WhatsApp or a similar, mobile telephone-based, bandwidth light app. While the end goal would be to ensure that all stakeholders have equal access to interoperable information systems, until that outcome is a reality, international and other responding stakeholders must be prepared to use the information technology that is available to and used by the affected community.

Our findings suggest that information sharing is a key area in which localization of humanitarian action is maturing albeit with areas where improvements can be made. As the world confronts a future wherein climate change-influenced hazards exacerbate risks and vulnerabilities, ensuring that stakeholders have access to information can help build resilience. Intentionally sharing information as broadly and swiftly as possible can mean that communities who are or will be impacted by disasters understand their vulnerabilities and can implement locally-developed programs to build capacity against those hazards with or without national, regional, or international input (beyond information, of course). A key limitation of our study is a predominance of regional and international practitioners whose perspectives may be overly rosy. Indeed, local stakeholders may see other avenues for improving information flows that are either already in practice or should be considered.

15.5.1 Future Studies

As part of an effort to round out these findings, there is scope for further research, in particular: (1) additional interviews that focus on the ways local actors prefer to communicate/receive information; (2) how localizing information sharing changes during complex emergencies; and (3) the applicability of our findings to geographic regions beyond the Asia–Pacific.

Abbreviations and Acronyms

AADMER	ASEAN agreement on disaster management and emergency response
ASEAN	Association of Southeast Asian Nations
CCA	Climate change adaptation
COVID-19	Coronavirus disease 2019
DRR	Disaster risk reduction
GNDR	Global Network of Civil Society Organisations for Disaster Reduction
HA/DR	Humanitarian assistance/disaster relief
HFA	Hyogo framework for action
IASC	Inter-agency standing committee
IM	Information management
IT	Information technology
NGO	Non-governmental organization
OCHA	Office for the coordination of humanitarian affairs (of the UN)
PIC	Pacific Island country/countries
SOHS	State of the humanitarian system
UN	United Nations
UNDRR	United Nations Office for Disaster Risk Reduction
UNISDR	United Nations Office for Disaster Risk Reduction (now UNDRR)

References

ALNAP (2022) The state of the humanitarian system. ALNAP Study. ALNAP/ODI, London

Alvesson M (2011) Interpreting interviews. Sage, London
Arikunto S (2010) Prosedur Penelitian (Suatu Pendekatan Praktik). Rineka Cipta, Jakarta
Ayobi Y, Black A, Kenni L, Nakabea R, Sutton K (2017) Going local: achieving a more appropriate and fit-for-purpose humanitarian ecosystem in the Pacific. Centre for Humanitarian Leadership, Fiji National University, and Humanitarian Advisory Group. https://humanitarianadvisorygroup.org/insight/going-local-achieving-a-more-appropriate-and-fit-for-purpose-humanitarian-ecosystem-in-the-pacific/
Barbelet V, Bryant J, Willitts-King B (2020) All eyes are on local actors: Covid-19 and local humanitarian action. Humanitarian Policy Group, Overseas Development Institute. https://www.odi.org/publications/17173-all-eyes-are-local-actors-covid-19-and-local-humanitarian-action-opportunities-systemic-change
Bonis-Charancle J, Vielajus M (n.d.) Aid localisation: current state of the debate and potential impacts of the Covid-19 crisis. Human Alternat https://www.alternatives-humanitaires.org/en/2020/07/23/aid-localisation-current-state-of-the-debate-and-potential-impacts-of-the-covid-19-crisis/
Carlson CM (2017) The revival and integration of traditional knowledge and practices for cyclones into government of the cook islands policies and planning. Master of Philosophy (MPhil) Thesis, Auckland University of Technology, Faculty of Culture and Science. https://openrepository.aut.ac.nz/bitstream/handle/10292/11743/CarlsonC.pdf?sequence=5&isAllowed=y
Cooley L, Gilson J, Ahluwalia I (2021) Perspectives on localization. Professional Services Council. https://safe.menlosecurity.com/doc/docview/viewer/docN4E5CFA8F8F013c7e3b4414916544c35db65a1b1ee97637395a04b1bbebea04b3818e7f258fed
Creswell JW (2011) Educational research: planning, conducting, and evaluating quantitative and qualitative research. Pearson Education International, New Jersey
Dekens J (2007) Local knowledge for disaster preparedness: a literature review. International Centre for Integrated Mountain Development (ICIMOD), Kathmandu, Nepal. https://www.preventionweb.net/files/2693_icimod8fc84ee621cad6e77e083486ba6f9cdb.pdf
Edmondson AC, McManus SE (2007) Methodological fit in management and research field. Acad Manag Rev 32(4):1155–1179
GNDR (Global Network of Civil Society Organisations for Disaster Reduction) (2020) Views from the frontline global report: why are people still losing their lives and livelihoods to disasters? Views from the frontline 2019. https://www.gndr.org/resource/views-from-the-frontline/views-from-the-frontline-global-report/
HAG, GLOW, and PIANGO (2022) Greening the system: a vision for a green humanitarian future. Humanitarian Horizons. HAG, Melbourne
Hyogo Framework for Action (2022) Prevention web. https://www.preventionweb.net/sendai-framework/Hyogo-Framework-for-Action. Accessed on 21 Oct 2022
IASC (Inter-Agency Standing Committee) (n.d.) About the Grand Bargain. https://interagencystandingcommittee.org/about-the-grand-bargain. Accessed on 7 Sep 2022
IDNDR (International Decade for Natural Disaster Reduction) (1994) Yokohama strategy and plan of action for a safer world: guidelines for natural disaster prevention, preparedness and mitigation. In: World conference on natural disaster reduction; Yokohama, Japan, 23–27 May 1994. UN Department of Humanitarian Affairs, Geneva, Switzerland. https://www.preventionweb.net/files/8241_doc6841contenido1.pdf
Kelman I (2013) Disaster mitigation is cost effective. World Bank, Washington, DC. https://openknowledge.worldbank.org/handle/10986/16341
Miles MB, Huberman AM (1994) Qualitative data analysis: an expanded sourcebook. Sage Publications, Thousand Oaks, CA
Schwarz GM, Stensaker I (2014) Time to take off the theoretical straightjacket and (Re-)introduce phenomenon-driven research. J Appl Behav Sci 50(4):478–501
UNDRR (UN Office for Disaster Risk Reduction) (2019) Words into action: local disaster risk reduction and resilience strategies. Geneva. https://www.undrr.org/publication/words-action-guidelines-implementation-guide-local-disaster-risk-reduction-and

UNDRR (UN Office for Disaster Risk Reduction) (2021) Words into action: nature-based solutions for disaster risk reduction. Geneva. https://www.undrr.org/words-action-nature-based-solutions-disaster-risk-reduction

UNDRR (UN Office for Disaster Risk Reduction) (2022) DRAFT–words into action: using traditional knowledges for disaster risk reduction. Geneva, Switzerland. https://www.undrr.org/publication/words-action-guidelines-using-traditional-knowledges-disaster-risk-reduction-public

UNDRR (UN Office for Disaster Risk Reduction) (n.d.) What is the Sendai framework for disaster risk reduction? https://www.undrr.org/implementing-sendai-framework/what-sendai-framework. Accessed on 21 Oct 2022

UNISDR (UN Office for Disaster Risk Reduction) (2013) Synthesis report: consultations on a post-2015 framework on disaster risk reduction (HFA2). Geneva, Switzerland. https://www.preventionweb.net/files/32535_hfasynthesisreportfinal.pdf

UNISDR (UN Office for Disaster Risk Reduction) (2015) Chart of the sendai framework for disaster risk reduction 2015–2030. https://www.preventionweb.net/files/44983_sendaiframeworkchart.pdf

Von Krogh G, Rossi-Lamastra C, Haefliger S (2012) Phenomenon-based research in management and organisation science: when is it rigorous and does it matter. Long Range Plan 45(4):277–298

Chapter 16
An Effective Business Continuity Planning and Financing

Takahiro Ono

Abstract Climate change is making meteorological disasters more serious, and complex and cascading disasters are causing great human and economic damage. New risks such as COVID-19 and cyberattacks are emerging. In such a society, complex and cascading unpredictable disasters are likely to occur in the near future. Currently, countries, regions and companies have made scenario-based preparations based on experience, but a series of events made it difficult for them to respond if they exceeded expectations. In order to respond to such unexpected events, it is desirable to develop response procedures and strategies for each management resource that becomes unavailable based on the outcome of the event rather than the cause of the event. This all-hazards approach is discussed in the chapter. It also suggests the effects of prefinance in risk management and a new style of BCP.

Keywords All-hazard approach · BCP · Business continuity · Consequence based · Risk finance · Risk management

16.1 Introduction

Based on the lessons learned from the major disasters that occurred, the disaster management system has been strengthened in Japan. Economic development and the interconnectivity of supply chains have raised concerns about economic losses due to indirect damage, and it has become necessary to enhance the effectiveness of business continuity plans (BCP) that have been introduced. New and unpredictable risks have emerged, such as COVID-19 and cyberattacks.

This chapter explains the limitations of scenario-based preparedness from past events and the all-hazards approach based on consequential events to deal with unexpected events. It also suggests the effects of prefinance in risk management and a new style of BCP.

T. Ono (✉)
Tokio Marine Holdings, Inc., Tokio Marine Resilience Co., Ltd., Tokyo, Japan
e-mail: takahiro.ono@tokiomarinehd.com

T. Izumi et al. (eds.), *All-Hazards Approach*, Disaster Risk Reduction,
https://doi.org/10.1007/978-981-97-1860-3_16

16.2 Comprehensive Disaster Management System

Japan belongs to the Pacific Ring of Fire and is frequently hit by earthquakes and volcanic eruptions. About 80% of the world's largest earthquakes have occurred in the region, including several in Japan. Tectonic plates around Japan move slightly every year, accumulating stress that will cause tsunamis when the energy is released. Because of its geographical, topographical, and meteorological conditions, Japan is subject to frequent natural hazards such as typhoons, torrential rains, and heavy snowfall. Until the mid-1950s, numerous large typhoons and earthquakes caused extensive damage and thousands of casualties (Fig. 16.1).

In 1959, the Ise Bay Typhoon occurred, killing more than 5,000 people. Reflecting on the damage caused by this catastrophe, the government began building a comprehensive disaster management system (Cabinet Office 2018b). In 1961, the Disaster Countermeasures Basic Act was adopted and the Central Disaster Management Council, chaired by the Prime Minister, was established to develop a strategic disaster management system.

Legal systems are being reviewed in light of the lessons learned from major disasters. For example, the government created earthquake insurance after the 1964 Niigata earthquake, revised the disaster relief fund system for heavy rains in 1967, and revised the Building Standards Act after the 1978 Miyagi earthquake. The country has developed disaster management systems, promoted national land conservation, improved weather-forecasting technologies, and upgraded disaster information communication systems. As a result, the number of deaths caused by natural disasters declined from the 1940s to the 1980s (Fig. 16.1).

Despite such efforts, however, more than 6,400 people died in 1995 in the Great Hanshin–Awaji earthquake (the Kobe earthquake). In 2011, the Great East Japan earthquake and tsunami killed more than 20,000 people. These two major natural disasters offer many insights and lessons to be learned.

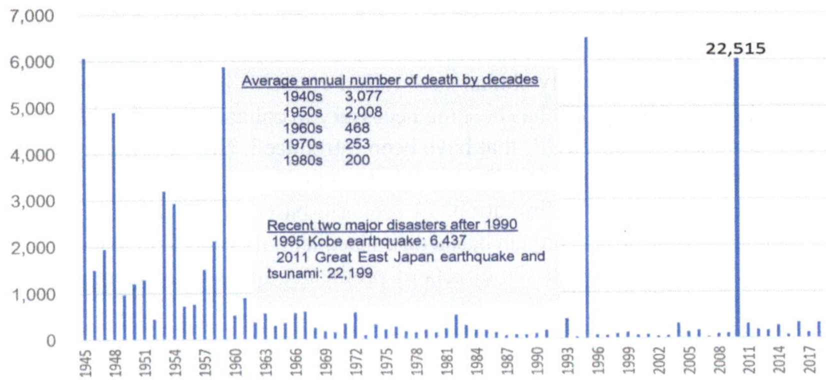

Fig. 16.1 Casualties due to natural disasters. *Source* Made by author

In order to avoid repeating the same mistakes in the future, all stakeholders in society continue to work toward resilience and innovation.

16.3 Increased Economic Damage

16.3.1 What is Business Continuity?

When disasters strike, the supply of goods and services provided by companies may be disrupted. Prolonged interruptions could have an impact on a company's finances and, in the worst-case scenario, even bankruptcy. Users and customers of the company's services are also affected by the lack of availability. If the outage is prolonged, customers will switch to other services providers and not come back, severely affecting the company. Therefore, it is necessary to know who would be affected and to what extent if the goods or services provided by the company were interrupted. It is also necessary to consider not only the financial impact of the company, but also the impact on customers and the social valuation of the company.

It is also extremely important to have a preliminary assessment of the maximum tolerable period of disruption (MTPD) for each product or service (Fig. 16.2).

With this in mind, various measures must be taken to ensure that products and services do not exceed the MTPD. In ISO 22301 (2019), Security and Resilience–Business Continuity Management Systems–Requirements, the "Business Continuity" and "BCP" are defined as follows. "Business continuity is the capability of an organization to continue delivering products and services within acceptable time-frames at predefined capability during a disruption." And "BCP is documented information that guides an organization in responding to a disruption and in resuming,

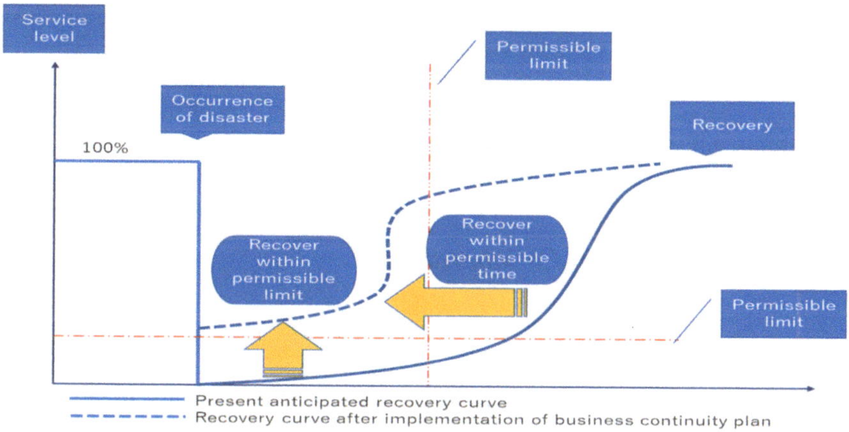

Fig. 16.2 Concept of business continuity. *Source* Made by author

recovering, and restoring the delivery of products and services consistent with its business continuity objectives" (ISO 2019).

It is also important to prioritize which products and services will start to be supplied when resuming. These business continuity concepts should apply not only to a single private company, but also to public agencies, multiple organizations, and even regions. In the case of a region, it is desirable to understand the extent of the impact of interruptions in social functions such as electricity, telecommunications, logistics, finance and medical care, etc. It would then be beneficial to introduce preventive measures to restore each social function within the MTPD by the interruption (Ono 2015).

16.3.2 Dissemination of Business Continuity Concept

The government assumes that the Tokyo metropolitan area will soon be hit by a major earthquake and has taken several measures to mitigate the expected massive human and property damage. In order to reduce economic damage, the government has set a target of 100% for large-sized companies and 50% for medium-sized companies to adopt BCP in the private sector, and has been promoting activities to spread and raise awareness of BCP, including the publication of the Business Continuity Guidelines (Cabinet Office 2014a). These efforts have been successful, and companies are working to improve their BCPs.

As a result of this initiative, a number of guidelines have been issued by local governments, chambers of commerce and industry associations, and the recognition of the concept of business continuity in the industry has increased dramatically, as reflected in the results of a survey conducted by the Cabinet Office on the implementation status of BCPs (Cabinet Office 2014b, 2016, 2018a). In 2007, about 20% of large-sized companies and 10% of medium-sized companies had a business continuity plan, but by 2017 this had increased to more than 60% of large-sized companies and about 30% of medium-sized companies (Figs. 16.3 and 16.4). The first version of the international standard ISO 22301 (2019), "Security and Resilience–Business Continuity Management Systems–Requirements" published in 2012, may have further influenced this growing interest.

16.3.3 Importance of Collaboration

While this concept is becoming more common, the two catastrophes of 2011 forced companies to take a deeper look at their risk management. One is coordination between supply chains and the other is public–private partnerships in the region. In the Great East Japan Earthquake, many lives were lost and business activities were disrupted. According to a survey by Teikoku Databank (2016), 650 private companies with about 10,000 employees went bankrupt within a year of the disaster.

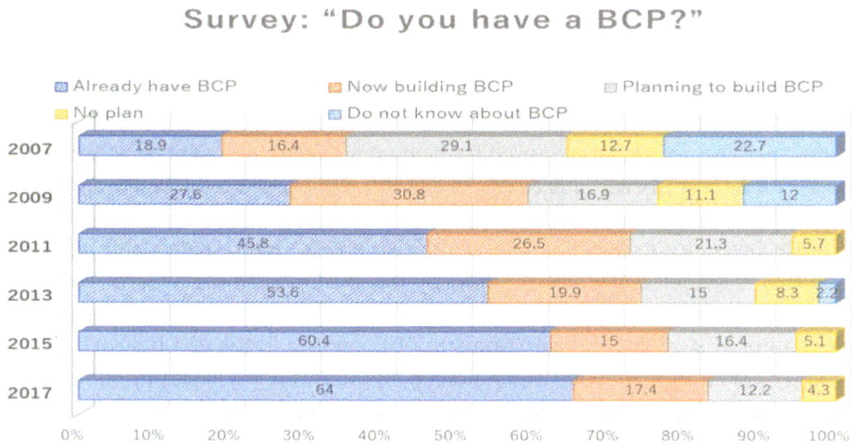

Fig. 16.3 BCP status of large-sized companies. *Source* Made by author

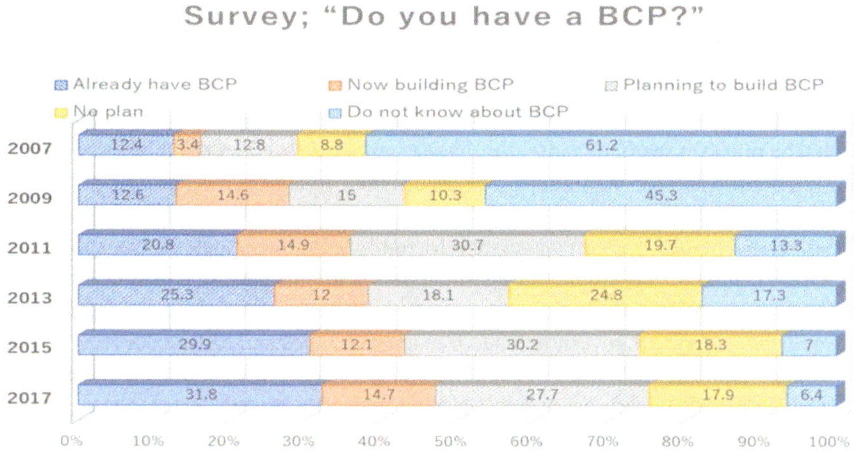

Fig. 16.4 BCP status of medium-sized companies. *Source* Made by author

However, only 101 companies, or 15% of the total, are located at the tsunami affected region. Many of the companies that went bankrupt were small businesses spread across the country and suffered "indirect losses". All, including small and medium-sized enterprises, recognized the need to incorporate supply chain and indirect loss concerns into BCP. With regard to the flooding in Thailand that occurred in the autumn of the same year, it was required to comprehensively assess the actual risks surrounding suppliers and subcontractors in order to recognize the risk of supply chain disruption, as well as flood control measures at their own sites such as the maintenance of drainage pumps.

Companies have naturally realized cost savings by enhancing the efficiency of their supply chains, but the recent disaster has shown once again that a partial supply chain outage can have a negative impact on the entire supply chain, and that the more efficient the supply chain, the more vulnerable it is to disasters. In addition to the "just-in-time" approach used by many manufacturing companies, we recognized that a "just-in-case" approach was useful and needed to be better combined. If you don't have an alternative source of parts or services, you may also need to increase inventory to avoid supply chain disruptions and to understand the status of upstream suppliers who are not directly involved in business with the company. There are diamond and pyramid structures in the supply chain, depending on the availability of such alternatives (Ministry of Economy, Trade and Industry (Japan) (METI) 2019). In particular, when a company in the supply chain has proprietary technology with a high global share, it is necessary to analyze and visualize whether its supply chain is a "diamond structure" or a "pyramid structure" and take necessary measures (Fig. 16.5).

In this way, "efficiency to eliminate unnecessary costs" and "redundancy or substitution" seem to be diametrically opposed rationales, but in reality, it is argued that the organic integration of the two in the process of analyzing and strengthening supply chains can improve the business environment and enhance competitiveness.

In the Great East Japan Earthquake, many local governments that were supposed to protect their communities were damaged and lost their functions, interrupting the initial response immediately after the disaster and the subsequent processes necessary for restoration and reconstruction. The loss of their functions has had a major impact on the activities of private companies and the daily lives of residents. On the other hand, many of the lifelines necessary for daily activities, such as transportation, electricity, gas, telecommunications, hospitals and supermarkets, are operated by the private sector. In addition, many private companies have signed disaster relief agreements with local governments to transport supplies and restore damaged roads. The responsibilities and roles of stakeholders in civil society are closely intertwined and interdependent with economic development. The private sector plays a major role in creating jobs, supporting local economies and ensuring the sustainability of local communities at normal time. And this role becomes even more important in the

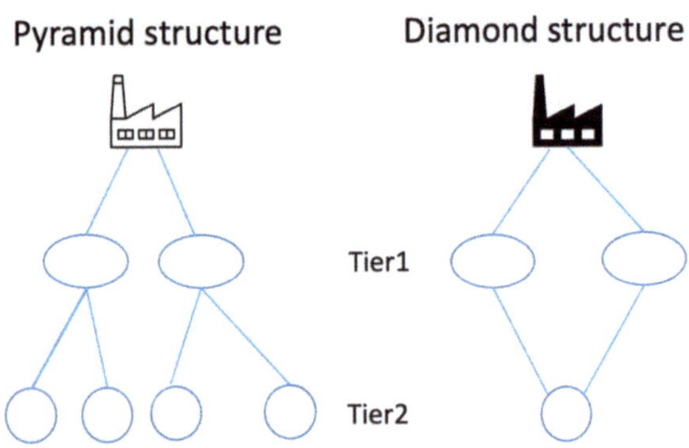

Fig. 16.5 Types of supply chain structure. *Source* Made by author

event of a disaster, as they have more resources and capital to provide relief supplies and evacuation space to affected employees, business partners, and residents. We were reminded that in order for private companies to fulfill their required roles and functions to contribute to maintaining the resilience of society, companies themselves need to become more resilient to disasters. And local governments must create an environment to support them (Ono and Anbumozhi 2021).

As a result of these experiences, more and more companies are incorporating the perspective of collaboration with local governments and communities into their business continuity initiatives that have already been formulated. Others regret the lack of sufficient training and unsatisfactory response, and the failure to fully review and inform employees about the content of the BCPs that was developed. The effectiveness of BCPs must always be ensured through periodic drills and ongoing training.

16.3.4 Finance and Risk Management

This section discusses the effectiveness of pre-financing investments in risk management. The UNDRR (United Nations 2022) says that "Every USD 1invested in risk reduction can save up to USD15 in post disaster recovery," and the World Bank notes "that Every USD1 invested in making infrastructure disaster resilient saves USD4 through fewer disruption and reduced economic impacts".

In the process of implementing risk management (Fig. 16.6), it is necessary to evaluate the amount of risk identified and consider reducing the amount of risk, when necessary. Mitigating the amount of risk can range from immediate to large investments. These countermeasures can be classified into hard and soft countermeasures.

For example, increasing IT server capacity and building backup sites are hard measures. There are also soft measures such as rulemaking and action plans to prevent accidents during normal times, and advance training to prevent the spread of damage when dangers become apparent. Risk mitigation also includes ways to reduce the likelihood of a risk occurring and to reduce the magnitude of the impact when it does occur.

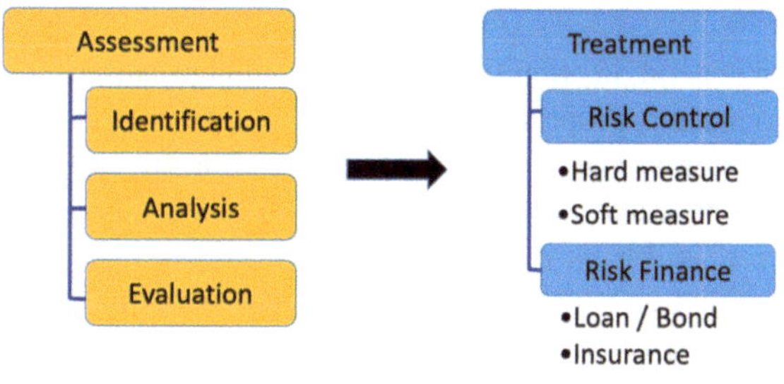

Fig. 16.6 Process of risk management. *Source* Made by author

Therefore, it is important to determine what measures should be introduced for mitigation, taking into consideration the direction of effectiveness (i.e., whether to reduce the frequency of occurrence or the degree of impact), cost-effectiveness, and the possibility of introducing such measures over a short or long period of time. Thus, while introducing effective measures reduces risk, implementation of measures requires a certain level of investment.

While effective ex-ante investments to introduce these measures contribute to a reduction in the amount of risk, relying solely on reactive measures without ex-ante measures results in greater losses. This is exactly what Priority 3 of the Sendai Framework for Disaster Risk Reduction (SFDRR), "Invest in Disaster Risk Reduction for Resilience", advocates (United Nations 2015). As mentioned above, after introducing various measures to reduce the risk amount, the next consideration is whether the residual risk amount is within the acceptable range. When it is within the acceptable range, the risk should be taken. If the risk exceeds acceptable limits, the risk should be transferred through insurance or other risk financing mechanisms.

While many companies see insurance as an important source of recovery funds, they should recognize that insurance is not suitable for all funding needs after the disaster. Costs to prevent loss or recurrence are reimbursed in advance and are generally not covered by insurance. Risk management should be implemented in line with these policies, which require accurate assessment of risks and reduction of risk amount through appropriate risk control measures. It should be emphasized that the best mix of the effects of pre-investment and the hedging of residual risk by insurance and other mechanisms is important.

It is also necessary to consider indirect as well as direct damage when assessing the amount of risk. The importance of a risk finance perspective when considering a BCP was discussed in the previous section. Direct damage is mainly to people and property. The indirect damages are loss of profits and loss of sales due to business interruption. With the increasing complexity of supply and value chains in recent years, this is an important aspect that could cause significant economic damage.

16.4 Towards Resilient Society

16.4.1 Robustness to Resilience

In general, in order to mitigate the damage caused by natural disasters, it is necessary to make appropriate preparations in advance to narrow the scope of negative effects, shorten the period of disruption, and at the same time, strengthen organizational, physical, and behavioral skills, and acquire appropriate skills to respond to emergencies. In this way, society, business activities and daily livelihood can be continued without interruption, and even if there is an interruption, conditions can be quickly restored to normal. This is a fact that applies at the national level, at the single-organization level, and at the individual level. However, the methods of disaster prevention and mitigation measures promoted by governments and organizations to mitigate the damage and impact of disasters need to be examined according to the maturity level of governments and organizations and existing rules and systems.

Here we consider the concept of resilience, which has been much discussed in recent years. The definition of resilience by UNDRR and the World Economic Forum (WEF) is as follows:

UNDRR: The ability of a system, community or society exposed to hazards to resist, absorb, accommodate, adapt to, transform and recover from the effects of a hazard in a timely and efficient manner, including through the preservation and restoration of its essential basic structures and functions through risk management (UNDRR, 2022).

WEF: The ability of an organization to overcome external shocks and grasp new opportunities in their wake (World Economic Forum 2022).

Interestingly, both definitions refer to the "ability" to not only overcome and adapt to challenges faced, but also to seize new opportunities. Although the concept of resilience has become popular, it is important to remember that some robustness is still needed. It is also important to consider the following time series in order to examine the relationship between resilience and robustness and to consider optimization.

In the event of a disaster, initial response activities are needed to mitigate the damage to lives and property. To carry out these activities meaningfully, it is important to collect information promptly and make appropriate judgments, and it is desirable to prepare procedures in advance. Initial response activities need to be considered separately for each hazard because of the different characteristics of earthquakes and typhoons. In addition to initial response activities, organizations should also consider their strategies for dealing with business interruptions. As noted above, organizations should develop a BCP before an event to minimize the negative impact of interruptions in the supply of products and services. Therefore, in many cases, when a disaster strikes, initial response activities begin first and business continuity activities begin a little later. Generally, depending on the hazard, first response activities are conducted by members of the facility management team, while business continuity activities are often discussed by members of the strategic division team. These two groups of activities are often fragmented. Therefore, it is necessary to pay attention to smooth coordination between initial response activities and business continuity activities.

Next, the balance between the two aspects of "preparedness" and "response" after an incident is also critical. Preparedness refers to preparations made in advance and can be categorized into two aspects: preparations during normal times to prevent the occurrence of an incident, and preparations to appropriately respond to an incident once an incident has occurred. In order to respond appropriately, it is beneficial to have manuals ready and capacity building such as education and training. The same applies to business continuity activities. Preparation prior to an incident is to consider strategies and measures to minimize the impact of a business interruption, and response to an incident is to take action to prevent the impact of the business interruption from being magnified. Thus, taking measures in advance to prevent an occurrence of an incident leads to "robustness," while taking appropriate measures after an incident occurs is related to "resilience." Therefore, excessive reliance on one or the other is very dangerous, and it is important to strike a balance between the two. Consideration of the aspects of "initial response" and "business continuity," "preparedness" and "response" is important in this process.

16.4.2 Consequence Based Design

In response to the unpredictable VUCA era, in which VUCA is a collective term for Volatility, Uncertainty, Complexity, and Ambiguity, the concept of an "all-hazards approach" (sometimes called multi-hazards) is being discussed. In such an environment, there is a growing recognition that when considering responses to events, it is appropriate to focus on the consequences rather than the cause of the event and take measures.

So far, we have assumed the damage from past experiences and case studies and considered measures to be taken in the event of an occurrence. Manuals and BCPs, for example, are often written based on certain assumptions based on past experience, such as how long critical infrastructure is down and how many days it takes to restore buildings and facilities. However, there are obviously many cases where expectations are not met, and such cases are often described as "unexpected".

The all-hazards approach, on the other hand, can be translated into the idea that whatever event occurs, it is ultimately functions and management resources that are affected. In this way, the all-hazard approach will consider measures at each level on the assumption that functions and resources will become unavailable regardless of the cause. There are multiple options for each failure level of management resources, such as waiting for recovery if the outage lasts for 24 h, running the in-house power generation if the outage lasts more than three days, or moving to another location if the outage lasts more than a week.

Previously, BCP assumed only one combination of situations in which management resources were unavailable due to a single hazard. Therefore, there is only one major option prepared per resource. If the event does not turn out as expected, the effectiveness will be reduced. However, by providing multiple options for measures for each resource, it is possible to flexibly combine measures for resources that are not available in accordance with the actual situation at the time, and thereby enabling to respond to all kinds of events. Until now, many BCPs had to be created for each hazard, including earthquake BCPs, pandemic BCPs, and typhoon BCPs. However, creating a BCP based on the all-hazards approach has the advantage of covering all hazards in one BCP.

It should be noted that "situational awareness" and "combining countermeasure options" require skills and decision-making. When an incident occurs, decision-making skills are necessary to act appropriately, and decision-making requires information. It is more important than ever to build and strengthen organizational structures and mechanisms that can collect information, analyze collected information, and make appropriate decisions. To overcome this, education and training are needed to ensure that employees can act quickly and properly in an emergency.

16.4.3 Back to New Normal

In 2019, the Covid-19 spread rapidly around the world. In Japan, a state of emergency has been declared and restrictions on travel and leaving home have been implemented to control the spread of the infection. The Japanese government requested companies to reduce the number of employees who come to office by 70%, and to reduce traffic congestion by changing commuting times (Tokyo Chamber of Commerce and Industry 2020). The government also enhanced the companies' efforts by disclosing the names of the companies and their implementation status.

The use of online tools such as ZOOM, TEAMS, and WebEx has accelerated to support the use of remote work. As this remote work environment is being created, more and more people are finding creative ways to perform various tasks that previously could not be done without coming to the office. For example, internal approval procedures using stamps and approval forms have been replaced by electronic approval systems, and fax communications have been replaced by electronic data of PDF files and the use of e-mail. It has also enhanced its online decision-making capabilities by allowing people to participate in meetings online, which previously would have been considered absent if they were not in the room. It was initially introduced as a temporary interim measure, but in the years since it went into operation, a growing number of organizations have reconsidered their business processes and changed their rules, shifting to regular implementation rather than reverting to traditional practices.

What follows describes how this situation affects an organization's BCP. In developing the BCP, it is necessary to consider what to do when the management resources necessary for operations run out. It has been considered effective to prepare a backup office in case an office becomes unavailable, such as in the event of an earthquake. The practice of switching to a backup office was then often conducted. Covid-19's push for remote work through telecommuting seems to have lowered the need and hurdle for switching to a backup office. In other words, the addition of the option to work remotely from home, to the original alternative solution of having backups in a separate venue, has increased the level of existing BCP and improved the resilience of the organization.

As Covid-19 converges and the telecommuting model normalizes at some level in the future, the current temporary remote structure will change to normal. Previously, telecommuting and remote work were considered temporary responses, but they are breaking common sense and becoming part of a new approach. Instead of sticking to common sense, we should recognize the need to constantly think about innovative ideas. In the past, recovery strategies and measures have been considered on the assumption that customer and market needs will continue to exist. This is a common method of Back to Normal and existing BCPs are also built around this idea.

However, as is well known, behavioral changes and market/demand fluctuations are undergoing profound transformation. We need to keep a close eye on these trends and developments. Since the end of Covid-19 won't bring us back to where we were before Covid-19, it's important to consider strategy with Back to New Normal in mind. Furthermore, we believe this will be a steppingstone for innovation.

Next, it is important to build a flexible system that can seamlessly integrate normal and emergency activities. In the past, for example, when a disaster occurred, members would gather at the headquarter buildings where the emergency operation center was located. But with online resources available, there's no need to convene at headquarters.

In the early days of Covid-19, employees with limited IT skills couldn't connect even if they wanted to, but now everyone is used to the system and usually works remotely. In this context, it has become recognized that processes such as "decision making" that were previously considered impossible are fully possible online, and that innovation can occur through the expansion of those processes and capabilities. This is truly a paradigm shift, and we must remember that in the age of VUCA, we need to be constantly searching for better ways of doing things now more than ever.

References

Cabinet Office (Japan) (2014a) Business continuity planning guideline

Cabinet Office (Japan) (2014b) Survey on BCP status in Japan 2013

Cabinet Office (Japan) (2016) Survey on BCP status in Japan 2015

Cabinet Office (Japan) (2018a) Survey on BCP status in Japan 2017

Cabinet Office (Japan) (2018b) White paper on disaster management 2018

ISO (2019) ISO22301, Security and resilience—business continuity management systems—requirements

Ministry of Economy, Trade and Industry (Japan) (METI) (2019) White paper on manufacturing industries 2019

Ono (2015) International cooperation and disaster management, Chapter 6 'Towards community resilience and private partnership against hazards in Asia.' Kyoto University Press, Kyoto (in Japanese)

Ono T, Anbumozhi V (2021) Effects of business continuity planning on reducing economic loss due to natural disasters. Economic Research Institute for ASEAN and East Asia (ERIA) Discussion Paper Series No.350, 2020

Teikoku Databank (2016) Trend survey of 'Bankruptcy related to the Great East Japan Earthquake' (cumulative for 5 years after the outbreak). https://www.tdb.co.jp/report/watching/press/p160301.html

The Tokyo Chamber of Commerce and Industries Report 2020. https://www.tokyo-cci.or.jp/page.jsp?id=1022366

United Nations (2015) *Sendai framework for disaster risk reduction.* https://www.preventionweb.net/drr-framework/sendai-framework. Accessed on 30 April 2020

United Nations (2022) UNDRR website. https://www.undrr.org/terminology/resilience

World Economic Forum (2022) A framework for building organizational resilience in an uncertain future insight report

Chapter 17
National Disaster Management Agency (NDMA)'s Policy on National Disaster Response Force (NDRF) as All Hazard Response Force

Sujata Saunik and Alok Avasthy

17.1 Background

Before 2005, India depended on the armed forces to manage disaster. The vast subcontinent of India has been prone to the impact of climate change and climate vulnerability due to its long coastline, mountainous and hilly areas, and monsoon impact. Each time a natural disaster occurred anywhere in the country, it took much time to mobilize an effective response to deal with the emerging situation. Precious time was lost in doing civil-military liaison, transporting relief supplies and essential commodities to the affected area, and organizing search and rescue operations. Many a time, the responders reached the affected area more than 24–48 h after the event had occurred. In view of this, the Central Government at the highest level decided to constitute a national apex body named National Disaster Management Authority (NDMA) with the Prime Minister as its Chairman to lead disaster response for both natural and man-made disasters in the country. Soon enough, NDMA realized that there has to be a dedicated disaster response force that can be quickly deployed in disaster zones. And so, the National Disaster Response Force (NDRF) was set up as a single, dedicated, all-hazard approach force. The chapter explains in detail the evaluation of these organizations to provide the implementing/executing arm for disaster relief operations. This chapter shall also elaborate on the growth and activities of these two essential institutions for disaster management in India, explaining

S. Saunik (✉)
Government of Maharashtra, Maharashtra, India
e-mail: ssaunik@gmail.com

Takemi Fellow 2018, T.H. Chan School of Public Health, Harvard University, Cambridge, MA, US

A. Avasthy
Central Reserve Police Force, New Delhi, India

T. Izumi et al. (eds.), *All-Hazards Approach*, Disaster Risk Reduction,
https://doi.org/10.1007/978-981-97-1860-3_17

in detail also overseas international operations and the all-hazard approach which has developed over the years as the country's main plank for disaster response.

India is a vast subcontinent with different topography and climatic variations. As per the recent report of the Council for Energy, Environment and Water (CEEW), several states in India face extreme events of climate weather patterns where often the same state could be responding to two very different extreme events. This stretches the ability of the local administration to respond effectively to emerging crises. Often, they need outside support to be able to manage. Against this backdrop, along with NDMA, the State Disaster Management Agency (SDMA) was set up by the states. Also, upon seeing the success of several successful response operations led by NDRF, several states started setting up of the State Disaster Response Force (SDRF) one by one. The NDRF was trained to respond to all hazardous situations effectively.

India's distinctive geography, climate, and socio-economic conditions expose it to a range of natural and man-made disasters. Floods, droughts, cyclones, earthquakes, landslides, avalanches, and forest fires are some of the disasters that the country is highly susceptible to. Most of the land area, around 58.6%, is at risk of moderate to very high-intensity earthquakes, and floods and river erosion are prone to over 12% of the land or 40 million hectares. Cyclones and tsunamis threaten almost 5,700 km of the country's 7,516 km-long coastline. Additionally, nearly 68% of cultivable land is susceptible to drought, and hilly regions risk landslides and avalanches.[1] With the constant risk of industrial accidents and other man-made disasters like radiological, biological, and chemical hazards, managing disasters in India is a colossal undertaking.

In recent decades, the Indian government has undertaken significant initiatives to improve the country's disaster management approach. This marks a departure from a relief-focused strategy to a more inclusive approach encompassing all stages of the disaster management cycle, including prevention, mitigation, preparedness, response, relief, and rehabilitation. In December 2005, the government implemented the comprehensive Disaster Management Act, which establishes institutional mechanisms for creating and implementing disaster management plans, ensuring timely and effective responses to disasters by various levels of government, and reducing the impact of disasters. The Act establishes Disaster Management Authorities at the national, state, and district levels, as well as the specialized NDRF and National Institute of Disaster Management (NIDM) (Fig. 17.1).[2]

---

[1] Disaster management in India: Ministry of Home https://ndmindia.mha.gov.in/laws-plans-and-policies.

[2] Training Regime for Disaster Response (Volume – I): https://ndma.gov.in/sites/default/files/PDF/Reports/training-regime.pdf.

Fig. 17.1 Chronology of events

17.2 All-Hazards Approach

An all-hazards approach (AHA) is a comprehensive strategy designed to prepare and respond to any disaster, emergency, or crisis regardless of its cause, size, or complexity.[3,4] The approach emphasizes identifying, assessing, prioritizing, and addressing the potential risks and vulnerabilities in a community, organization, or country to effectively prepare, respond, mitigate, and recover from any hazard or threat.[5]

One of the fundamental principles of the AHA is risk assessment, which involves identifying potential hazards, examining their likelihood and potential impact, and prioritizing them based on their severity, frequency, and consequences.[6] This helps determine the resources, capabilities, and plans needed to address each hazard and develop a comprehensive emergency management plan that outlines roles, responsibilities, and procedures for responding to different types of disasters.

Another critical component of the AHA is preparedness, which involves developing and testing emergency plans, conducting training and exercises for emergency responders and community members, establishing communication systems, and ensuring the availability of necessary resources and supplies to respond to emergencies.[7] This helps to minimize the impact of disasters and to save lives and property by providing a timely and effective response.

The response is another crucial aspect of the AHA.[8] It involves mobilizing resources and emergency responders to the disaster site, coordinating the response efforts, and providing necessary services and support to the affected population.

[3] Emergency Preparedness Regulation–Clarifications on Definitions (n.d).

[4] Guide for All-Hazard Emergency Operations Planning (n.d.).

[5] Guide for All-Hazard Emergency Operations Planning (n.d.).

[6] Guide for All-Hazard Emergency Operations Planning (n.d.).

[7] Guide for All-Hazard Emergency Operations Planning (n.d.).

[8] Guide for All-Hazard Emergency Operations Planning (n.d.).

Fig. 17.2 All hazard approach

This involves working collaboratively with local, state, and federal agencies, non-governmental organizations, and the private sector to ensure a coordinated and effective response.

Recovery is the final stage of the AHA, which involves restoring normalcy and helping affected individuals and communities to rebuild and recover from the disaster.[9] Recovery efforts may include providing temporary shelter, medical care, and other essential services, repairing damaged infrastructure, and restoring economic activities (Fig. 17.2).

In recent years, the world has seen increased natural disasters, such as hurricanes, floods, earthquakes, and wildfires. These disasters can have a devastating impact on communities, both economically and socially. As a result, there has been a growing emphasis on an "all-hazard" approach to disaster management, which focuses on preparing for all types of disasters rather than just one.

One of the key benefits of the AHA is that it allows for greater efficiency and flexibility in disaster management. Rather than developing separate plans for each type of hazard, an AHA provides for a more coordinated response to disasters. This approach also allows for sharing resources and expertise between different agencies and organizations, which can be especially useful in large-scale disasters.

There is evidence to suggest that the AHA has been successful in reducing the impact of disasters. For example, a study by the National Institute of Standards and Technology (NIST) found that communities that had adopted an AHA to disaster management were better able to respond to Hurricane Katrina in 2005.[10] These communities had better communication networks, more effective evacuation plans, and more comprehensive emergency response plans.

Despite these successes, there are also challenges associated with the AHA. For example, it can be difficult to prioritize resources and allocate funding to effectively address all types of hazards. Additionally, competing priorities and interests may make it difficult to implement comprehensive disaster management plans.

The AHA is a comprehensive strategy that aims to prepare and respond to any disaster, emergency, or crisis. It emphasizes risk assessment, preparedness, response, and recovery and involves collaboration and coordination among all stakeholders to ensure a coordinated and effective response. By adopting this approach, communities, organizations, and countries can improve their capacity to respond to disasters and minimize their impact, thereby enhancing resilience and reducing the overall cost of disasters. The success statistics of the AHA in disaster management are promising. By preparing for all types of hazards, governments and communities can better protect themselves from the devastating effects of disasters. While there

[9] Guide for All-Hazard Emergency Operations Planning (n.d.).

[10] The Federal Response to Hurricane Katrina: Lessons Learned (2006).

are challenges associated with this approach, the evidence suggests that the benefits far outweigh the costs. As such, governments and communities must continue to prioritize the adoption of an AHA for disaster management.

An AHA disaster approach has been termed as an integrated approach for emergency preparedness planning which focuses on capabilities to handle the full range of emergencies and disasters by the United States center for Medicare and Medicaid Services. Globally, the AHA has been considered the principal framework for disaster managers. It is a highly effective framework for ensuring disaster planning and response (1). The US Federal Emergency Management Agency (FEMA) published a guide for AHA operation planning in 1996. This guide emphasized shared response planning for emergencies, which also formed the basis of the training of the NDRF in India. The NDRF has been trained in Chemical, Biological, Radiological, and Nuclear (CBRN) disaster management. This has helped the force effectively respond to the different types of incidents they have handled in the last decade.

17.3 Evolution of Disaster Management in India

The approach to disaster management in India has undergone significant changes. It has shifted from a reactive, activity-based system to a proactive, institutionalized structure. Additionally, it has moved from a single-domain system to a multi-stakeholder framework. Finally, there has been a change from a relief-focused approach to a comprehensive, proactive, and holistic approach to reducing risk in multiple dimensions ('multi-dimensional pro-active holistic approach for reducing risk'). India is a large country with different geography and typography. There is the Great Himalayan Region, a long coastline, desserts, and several remote and tribal areas where accessibility is an issue. As per the CEEW Report, India is vulnerable to the adverse effects of climate change and related disasters. Being the world's most populous country, the scale of both man-made and natural disasters due to climate-related events is enormous. The traditional Government systems of responding through Relief and Rehabilitation Departments soon became inadequate; with this realization, the NDMA and the NDRF worked together to prepare the disaster response, and mitigation plans through advocacy with the states, especially those whose hazard vulnerability index was high. And NDRF was designed to be an AHA response force. The four-pronged approach adopted was to plan for disasters through mitigation, preparedness, response, and recovery. This led to better preparedness for responding to different disasters rather than focusing on the most common or recurring events (Fig. 17.3).

A three-part disaster management system was established after the Disaster Management Act was implemented in 2005. The highest authority at the central level was the NDMA. In contrast, the state and district levels were managed by the State Disaster Management Authorities (SDMAs) and District Disaster Management Authorities (DDMAs), respectively. To handle hazardous disaster situations, Sect. 44(1) of the DM Act 2005 mandated the establishment of the NDRF. According

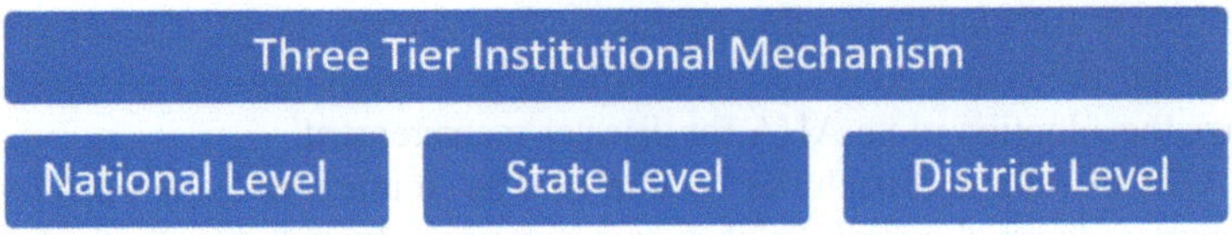

Fig. 17.3 The three tier structure

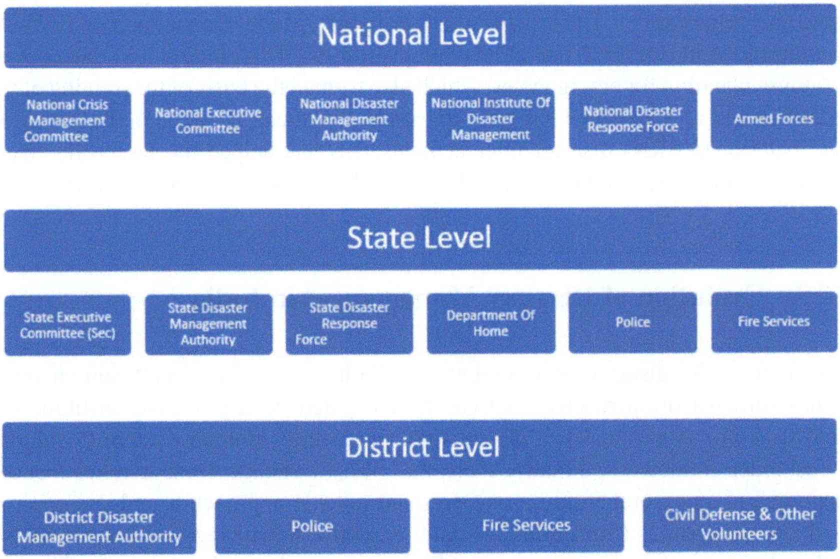

Fig. 17.4 National state and district levels structure

to Sect. 45 of the Act, the NDMA has general oversight, direction, and control over the force. At the same time, the Director General of NDRF, whom the Central Government appoints, is responsible for commanding and supervising the force. Verifying the claim that NDRF is currently under the MHA is necessary (Fig. 17.4).

17.3.1 NDMA

The Prime Minister of India heads the National Authority as the chairman. The vice-chair and members, drawn from various faculties like public health, floods, earthquakes, and other related fields, work with a team of research officers and officers drawn from various technical institutions and civil-military liaisons. This agency advises various state governments to prepare their disaster management and mitigation plans based on their hazard vulnerability risk assessment. Core groups have organized several guidelines headed by various members to guide disaster risk

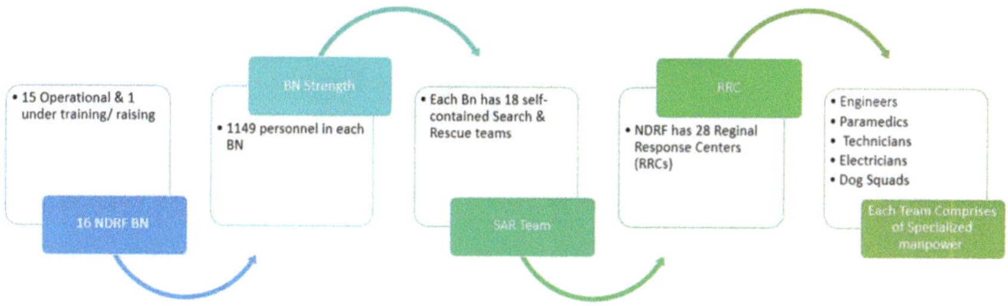

Fig. 17.5 Composition of NDRF

reduction and mitigation measures. According to the risk assessment, mock drills and emergency exercises are done periodically to test the response mechanism and efficacy of the emergency operation centers across the country.

The NDMA also prepared the policy for the training and equipment of the NDRF and a Standard Operating Procedure for their deployment. During peacetime, the NDRF works with communities in hazardous areas to build capacity and prepare the community to be the first responders before external help comes to their rescue.

17.3.2 Establishment of NDRF

India experienced a series of natural disasters between 1990 and 2004, which prompted the government to pass the Disaster Management Act of 2005. This was a significant step towards creating a three-tier institutional mechanism and implementing a comprehensive and integrated approach to disaster management. The act also significantly changed from the previous relief-centric response approach to a more proactive approach emphasizing prevention, mitigation, and preparedness. This approach aims to preserve developmental progress and minimize the loss of life, livelihoods, and property during disasters. Under this Act, the NDRF was formed to provide specialized search and rescue during and after disasters, and proactive deployment in impending disaster situations, and develop and build response capabilities of other state and local level stakeholders. On January 19, 2006, the NDRF was established as the country's top-tier search and rescue response force. Its motto is 'Aapada Seva Sadaiv Sarvatra,' which translates to 'continuous disaster response services under all circumstances'.[11] In 2016, the National Disaster Management Authority (NDMA) issued the first National Disaster Management Plan, which was reviewed and revised in 2019 (Fig. 17.5).

The NDRF currently consists of 16 battalions that will be strategically located as necessary. These battalions will collaborate closely with specified state governments and be readily available to assist them in any significant emergency circumstance.

[11] NDRF Raising Day: History, Significance, Facts And Role of National Disaster Response Force. https://www.ndma.gov.in/

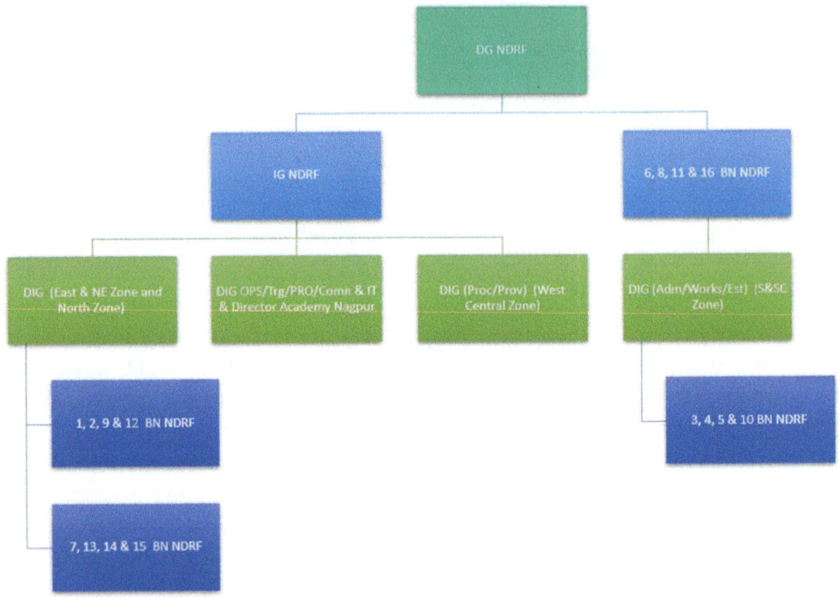

Fig. 17.6 NDRF–organization

While all NDRF battalions are responsible for managing natural disasters, four battalions will receive specialized training and equipment to handle emergencies related to CBRN events.

As the community is typically the initial responder during any disaster, NDRF is continuously involved in organizing "Community Capacity Building and Public Awareness and Preparedness Programs" for various organizations, such as the State Disaster Response Force (SDRF), State Police, Home Guards, Civil Defense, Fire Services, National Cadet Corps (NCC), National Service Scheme (NSS), Nehru Yuva Kendra Sangathan (NYKS), and Non-Governmental Organizations (NGOs), Students, Volunteers and other stakeholders. These programs aim to prepare these groups for emergencies until specialist forces arrive at the location. This out-of-the-box thinking of policymakers to engage the specialized force directly with the community has resulted in achieving the National Vision on Disaster Management in multiple ways by adopting villages, empowering youth, large-scale capacity building of the community, gender and disability sensitive approach, integration of technologies and promoting ingenuity in response.[12] In 2021, NDRF introduced a group of 100 female disaster combatants and rescuers. Moreover, each NDRF battalion is authorized to have 108 female responders (Fig. 17.6).

[12] National Disaster Response Force; https://www.ndrf.gov.in/about-us.

17.3.3 NDRF and AHA

In the initial years, NDRF followed a top hazards approach for disaster response in the country. The various operations that it led for search and rescue, which included building collapse, train accidents, cyclones, floods, mudslides, etc., made evident that the NDRF response team needs to be fully prepared to meet any contingency on site. After that, AHA was institutionalized with the compressive preparedness plan for all eventualities. This plan focused on bridging the critical gaps in both capacities and capabilities of the NDRF to be fully self-reliant and prepared to meet the complete spectrum of emergencies, including natural disasters that they responded to. This new approach yielded better and more significant results than the earlier response approach by reducing the time taken for deployment and search and rescue. The success of this approach then became a hallmark of NDRF operations, including overseas deployment in the future.

The NDRF has adopted an "All Hazard" approach in several circumstances.

Preparedness: An AHA ensures that the NDRF is prepared to respond to a wide range of disasters, including natural disasters (such as earthquakes, floods, and cyclones) as well as human-made disasters (such as chemical spills, terrorist attacks, and industrial accidents) (National Policy on Disaster Management n.d.). By considering all possible hazards, the NDRF can develop comprehensive plans and strategies to respond to any emergency effectively. As a part of preparedness, NDRF adopts the AHA on Public awareness and education about various types of hazards; it educates the public about general safety practices, early warning systems, evacuation procedures, and other essential information (National Policy on Disaster Management n.d.). This approach helps build a resilient society that can respond effectively to any disaster.

Risk Assessment: The initial phase of addressing disaster vulnerabilities involves conducting risk and vulnerability assessments in disaster-prone areas. This process includes hazard zonation, mapping, and vulnerability analysis (National Policy on Disaster Management n.d.).

Response: By adopting an AHA, the NDRF can be flexible and adaptable in responding to emerging or unforeseen hazards. Disasters can vary in nature and intensity, and new hazards can occur over time. A broad framework allows the NDRF to quickly assess and respond to evolving situations without being limited to a narrow set of predefined scenarios.

Recovery: The immediate aftermath of a disaster is characterized by disorder and chaos, with numerous urgent and significant tasks related to disaster response, rescue operations, and relief efforts. Effective coordination is crucial to ensure a well-organized disaster response, aid in recovery, and achieve successful disaster risk reduction, as recommended by the AHA. Coordination plays a vital role in enabling efficient communication, sharing of resources, and seamless integration of efforts to maximize the overall effectiveness of the response. Additionally, it aids in effectively distributing resources, avoids unnecessary repetition, tackles deficiencies in response endeavors, and promotes a shared emphasis on sustainable recovery and long-term mitigation of risks. Through coordination, it becomes possible to tailor the response appropriately to suit the disaster's socio-cultural, economic, and geographic circumstances. (Standard Operating Procedure on Effective Coordination and Corporation During Disaster Response n.d.).

The adaptation of the AHA by NDRF acknowledges the complexity and unpredictability of disasters and ensures that the force is equipped to handle a wide range of emergencies. By considering all hazards, the NDRF aims to enhance its preparedness, response capabilities, and overall effectiveness in safeguarding lives and minimizing the impact of disasters.

17.3.4 NDRF's Overseas Deployment

17.3.4.1 Fukushima Nuclear Disaster, Japan, 2011: Stellar Response Amidst Nuclear Leakage

In one of the most devastating disasters of the decade, the Japan Triple Disaster (2011), earthquake and after that tsunami and successive nuclear leakage at Fukushima nuclear power plant, a 46-member team of NDRF, led by the then Comdt. Shri Alok Avasthy, was deployed at Onagawa from 27 March to 07 April 2011 for the first ever overseas operation of NDRF since its inception. While most of the international search and rescue teams started returning from the affected area, fearing for their safety in the wake of the nuclear leakage, CBRN trained and equipped team of NDRF braving adverse weather conditions continued with their search and rescue operations and managed to retrieve seven bodies from under the debris, more than two weeks after the disaster occurred in the area. In addition, they found and turned in cash worth fifty million yen and other valuable items they came across. The NDRF team received high praise from the local authorities, media, and the general public for their commendable efforts.[13]

On April 5, 2011, the government of Japan conveyed its appreciation to the Indian NDRF team for their efforts. Those at a rehabilitation camp in Onagawa town hall expressed gratitude to the Indian team and shared their experiences with the group leader. The mayor of Onagawa was impressed by the team's evident commitment and integrity. Similarly, the mayor of Rifu Cho also thanked the Indian team for their assistance and support to the people of Miyagi, specifically to Onagawa.

During his visit to India on December 28, 2011, at the invitation of the India Council of World Affairs, the former Prime Minister of Japan, Mr. Yoshihiko Noda, mentioned the NDRF's professionalism, hard work, and dedication. He was inspired by these qualities and expressed his desire to build a strategic and global partnership based on the "Kizuna" bond of friendship between people. He stated that *"Under the leadership of Comdt. Alok Avasthy, the forty-six force members, showed such outstanding discipline in their work that no one thought it had been the force's*

[13] National Disaster Response Force (NDRF) (n.d.).

overseas operation even since its creation. I imagine how hard it was for NDRF members, coming from India to work under the bitter chill of TohoKu's blustery means in late March. Despite such harsh conditions in Onagawa, a port town of Miyagi prefecture severely damaged by the Tsunami, they listened to the needs of local residents and worked painstakingly with their hands to search for victims' bodies and their belongings from piles of rubble more than ten meters deep. Their dedication, whole heartedness and warm smiles touched the hearts of those who were grieved at the disaster and gave them immeasurable hope".

In 2013, His Imperial Majesty the Emperor of Japan, Akihito and Her Majesty the Empress Michiko felicitated Shri. Alok Avasthy, the team leader of NDRF deployed at Japan.

17.3.4.2 Nepal Earthquake, 2015

In April 2015, Nepal experienced a powerful earthquake measuring 7.8 on the Richter scale. This caused extensive damage to both life and infrastructure, and there was a concern that a humanitarian crisis was imminent. The earthquake was felt across Northern India as well. To assist with the aftermath, the NDRF sent 16 Urban Search and Rescue teams to Nepal, which included over 700 rescuers and 18 trained dogs equipped with the latest technology. The government of India directed this effort. The NDRF teams rescued 11 injured individuals and retrieved 133 bodies from the rubble. Additionally, they set up six medical camps and assisted 1,219 people. The NDRF also played a crucial role in distributing relief materials to those affected by the Nepal earthquake, with 1,176.571 tons of supplies delivered via rail, road, and air routes.[14]

17.3.4.3 Turkey-Syria Earthquake, 2023

The world has noticed India's swift action in response to the 7.8 magnitude earthquake in Turkey in February 2023. This response is evidence of the readiness and effectiveness of our rescue and relief teams. India was one of the first responders and fulfilled its commitment to humanity through 'Operation Dost.' 3 teams of NDRF consisting of 151 personnel and the dog squad were sent to the affected areas for search, rescue, and relief operations. India gained the respect and title of a 'Dost' from Turkey as the specialist forces working in adverse weather conditions and daunting situations did their best to aid the quake-ravaged nation.

[14] National Disaster Response Force (NDRF) (n.d.).

17.3.4.4 Exercises that NDRF Was Involved in

SAADMEx-2015

Recent natural disasters in South Asian countries, such as the earthquakes in Nepal and Afghanistan, have highlighted the importance of collaboration and cooperation among the nations regarding disaster preparedness. It has become evident that regional efforts are necessary, which involve using foreign civil aid and military resources. In 2011, the South Asian countries signed a Rapid Response to Natural Disasters agreement. The main objective of this agreement is to enhance individual national efforts and promote regional cooperation in responding to disasters. SAADMEx serves as a platform where the affected countries can review and discuss their disaster response plans and capabilities, with the involvement of national authorities and external multilateral organizations. The three-day SAADMEx event, which took place in 2015, focused on crucial aspects like emergency command and control, coordination, decision-making, and notifications and the integration of regional and international humanitarian aid.[15]

SCO Jt. Ex.—2019

In 2019, the Shanghai Cooperation Organization (SCO) coordinated a collaborative training session centered on urban earthquake search and rescue. This exercise was hosted by the NDRF and involved the participation of all SCO member countries, namely Russia, Tajikistan, China, Kyrgyzstan, Pakistan, Uzbekistan, Kazakhstan, and India. Observers from the International Search and Rescue Advisory Group (INSARAG) and the AHA Centre also participated in the event. Following the INSARAG guidelines, the simulation exercise spanned four days and focused on outlining the roles and responsibilities of Urban Search and Rescue (USAR) teams at national and international levels. The exercise incorporated various plans such as the utilization of the On-Site Operations Coordination Centre (OSOCC), the establishment of an Emergency Operations Centre (EOC)/Integrated Command Post (ICP)/ Response Cluster, collaboration with local authorities, and discussions on CM Coordination. Initially, the SCO members intended to make this joint exercise an annual occurrence. However, the last time subsequent events took place was in 2019.[16,17]

[15] South Asian Annual Disaster Management Exercise (SAADMEX) (2015).

[16] National Disaster Response Force (NDRF) (n.d.).

[17] Union Minister of Home Affairs Shri Amit Shah to inaugurate SCO Joint Urban Earthquake Search & Rescue Exercise (SCOJTEX)-2019 (n.d.).

BIMSTEC DMEx-2017, 2020

India has been hosting the BIMSTEC Centre for Weather and Climate (BCWC) since 2014, facilitating collaboration among BIMSTEC countries regarding weather and climate. They have also established a connection with the Tsunami Early Warning Centre, sharing data since 2006. This cooperation may expand to include Central Asian countries due to their vulnerability to weather events. With the support of the NDRF, India successfully organized two BIMSTEC Disaster Management Exercises in 2017 and 2020. These exercises involved various activities and garnered active participation from member countries and international agencies.[18]

17.4 Conclusion

The NDMA and NDRF have become synonymous with disaster management and response in India; they have developed a deep trust with the masses and the administration alike. The utilization of preemptive availability and pre-positioning of teams, foreseeing potential emergency scenarios facilitated by the widespread presence of its battalions, has effectively ensured the containment of damages. NDMA continues to guide and support the SDMAs in the country.

While NDRF has discharged its mandated role of response with a high degree of proficiency in many major disasters, it has significantly taken on itself many other responsibilities beyond its mandate, namely, capacity building of State Disaster Response Forces, community awareness programs, school safety program, management of animals in disasters and organizing mock exercises involving all stakeholders for different disasters. Notwithstanding it's positive features, there may still be some gaps that need a proper SWOT study of the force to enhance its efficiencies and capabilities.

References

Business aviation Maldives–private jet flight to Male (n.d.) The Bay of Bengal Initiative for Multi-Sectoral Technical and Economic Cooperation (BIMSTEC). https://bimstecdmex2020.in/

DM Act 2005 (n.d.) https://ndma.gov.in/sites/default/files/PDF/DM_act2005.pdf

Emergency Preparedness Regulation - Clarifications on Definitions (n.d.) Centers for Medicare & Medicaid Services (CMS) (cms.gov). https://www.cms.gov/Medicare/Provider-Enrollment-and-Certification/SurveyCertEmergPrep/Downloads/FAQ-Round-Four-Definitions.pdf

Guide for All-Hazard Emergency Operations Planning (n.d.) The federal emergency management agency (fema.gov). https://www.fema.gov/pdf/plan/slg101.pdf

Handbook on Disaster Management for Nodal Officers (n.d.) https://nidm.gov.in/PDF/pubs/Handbook_NodalOfficer.pdf

[18] Business aviation Maldives–private jet flight to Male (n.d.).

National Disaster Response Force (NDRF) (n.d.) The preparatory meeting of Shanghai cooperation organization (SCO) joint exercise on urban earthquake search & rescue (SCOJtEx)-2019 to be held in New Delhi on 01st & 2nd November, 2018. https://ndrf.gov.in/pressrelease/preparatory-meeting-shanghai-cooperation-organization-sco-joint-exercise-urban

National Disaster Response Force (NDRF) (n.d.) Chemical gas leakage at Vishakhapatnam 2020. https://www.ndrf.gov.in/operations/chemical-gas-leakage-vishakhapatnam-2020

National Disaster Response Force (NDRF) (n.d.) Radiation emergency management at mayapuri scrap market, New Delhi by NDRF–APR. 2010. https://www.ndrf.gov.in/operations/radiation-emergency-management-mayapuri-scrap-marketnew-delhi-ndrf-apr-2010

National Disaster Response Force (NDRF) (n.d.) Inland operations. https://ndrf.gov.in/inland-operations

National Disaster Response Force (NDRF) (n.d.) EPAL earthquake–Apr. 2015. https://www.ndrf.gov.in/operations/nepal-earthquake-apr-2015

National Disaster Response Force (NDRF) (n.d.) Home. https://www.ndrf.gov.in/

National Disaster Response Force (NDRF) (n.d.) Japan Tsunami–Mar. 2011. https://www.ndrf.gov.in/operations/japan-tsunami-mar-2011

National Disaster Response Force (NDRF) (n.d.) Amonia gas leakage–Jan. 2011. https://www.ndrf.gov.in/operations/amonia-gas-leakage-jan-2011

National Policy on Disaster Management(NPDM) (n.d.) https://www.mha.gov.in/sites/default/files/2022-08/NPDM-101209%5B1%5D.pdf

National Policy on Disaster Management (n.d.) mha.gov.in. https://www.mha.gov.in/sites/default/files/2022-08/NPDM-101209%5B1%5D.pdf

South Asian Annual Disaster Management Exercise (SAADMEx) (2015) PreventionWeb. https://www.preventionweb.net/event/south-asian-annual-disaster-management-exercise-saadmex-2015

Standard Operating Procedure on Effective Coordination and Corporation during disaster response (n.d.) ndrf.gov.in. https://ndrf.gov.in/sites/default/files/EFFECTIVE%20COORDINATION%20H.pdf

Union Minister of Home Affairs Shri Amit Shah to inaugurate SCO Joint Urban Earthquake Search & Rescue Exercise (SCOJTEX)-2019 (n.d.) https://pib.gov.in/newsite/PrintRelease.aspx?relid=194196

Chapter 18
All-Hazards Approach for a Resilient, Sustainable and Inclusive Society

Rajib Shaw

Abstract The world is becoming more complex and cascading, complex and systematic risk become more important over the years. We have seen new risks, and new types of hazards, not just natural hazards but biological, technological, radiological, nuclear, cyber etc. This has urged to think of all-hazards approach (AHA) as the core of complex risk management framework. There are two major arguments evolved for all hazards approach over time. One group of researchers argue the importance of AHA by citing public health hazards, cyber security hazards etc., especially for the critical infrastructures as well as individual level hazards. The other group of researchers argue the concept of top hazards, instead of AHA, and the core to that argument is hazards may be common, but the risks are context specific based on the exposures and vulnerability. Available AHA assessment and preparedness tools both at the individual level as well as infra or asset unit level (like school or hospital or other critical infra) are available, however, its penetration and usage has been still limited. In this chapter, four critical challenges are identified as risk perception, risk assessment, risk mitigation and governance. As postscript, three key points are mentioned as: (1) AHA as assessment and THA (Top Hazards Assessment) as the key to preparedness, (2) synergy at local level is the key issue of AHA, and (3) AHA needs whole government, whole society involvement.

Keywords All hazards approach (AHA) · Risk assessment · Top hazards approach (THA) · Sustainable and inclusive society · Adaptive governance

18.1 Introduction

The world is going through a dynamic and ever-changing risk landscape. The Global Risk Outlook of World Economic Forum has described different types of risks (environmental, technological, financial, geo-political, socio-economic etc.) over the last several years (from 2007 onward) in terms of likelihood and impacts. The analysis

R. Shaw (✉)
Graduate School of Media and Governance, Keio University, Fujisawa, Japan
e-mail: shaw@sfc.keio.ac.jp

T. Izumi et al. (eds.), *All-Hazards Approach*, Disaster Risk Reduction,
https://doi.org/10.1007/978-981-97-1860-3_18

shows two important points: (1) inter-relation of different types of risks, and (2) environmental risks (like disaster, climate change, extreme weather, biodiversity losses) and becoming more prominent risk in terms of both likelihood and impacts (WEF 2020). The same analysis of 2021 has pointed out emergence of infectious disease as the most prominent risk due to COVID-19 in the year 2020 (WEF 2021). However, two new risks emerged: digital power concentration and digital divide. While the world has focused on more digital connectivity (online education, health, shopping, work from home etc.), it has also created a digital divide. The divide is due to developed and developing country, urban and rural areas (with rural areas relatively poor connectivity) and age groups (youth and old age groups). While our world changed more towards digital transformation, cyber security has emerged as a new risk in the year 2022 (WEF 2022). The year 2023 has marked two new risks: energy crisis and cost of living as new risks, which is due to prolonged war between Russia and Ukraine (WEF 2023). Thus, within a span of three years, infectious disease, digital power concentration, digital divide, cyber security, energy crisis and cost of living have emerged in the new risk landscape. However, the risk of environmental risk like climate related hazards, extreme weather, biodiversity losses remain at the top.

Thus, the world is becoming more complex and cascading, complex and systematic risk become more important over the years. We have seen new risks, and new types of hazards, not just natural hazards but biological, technological, radiological, nuclear, cyber etc. This has urged to think of all-hazards approach (AHA) as the essential of complex risk management framework. In this chapter, some reflections are provided on the need of AHA to address resilient, sustainable and inclusive society.

18.2 Literature Review

Although the AHA has been mentioned and discussed in literatures almost forty years back, there are not that much academic work available on this topic. While the emergency response approaches shifted to a comprehensive approach, AHA was adopted, mostly in the United States emergency management system. The FEMA (Federal Emergency Management Agency) in 1996 released the first all-hazards emergency operation planning guidebook (FEMA 1996). The characteristic of the guidebook was to focus on state and local authorities and to integrate mitigation into response and recovery activities for all hazards. The focus also changed from hazard based to risk-based approach, which incorporated all hazards. Then, gradually over different major international events have urged for incorporation of new hazards, like after 2001 New York twin tower's terrorist attack, terrorism is incorporated. After 2011 Fukushima nuclear accident, NATECH (Natural Hazards Induced Technological Disasters) have been incorporated, and after 2020 COVID-19 pandemic, infectious disease is included. The Sendai Framework for Disaster Risk Reduction (SFDRR) has focused significantly on AHA, although several agencies still follow the concept of multi-hazards approach rather than AHA.

One of the two pioneer work done in early 2000s was by Vol Lubiz et al. (2008), which has argued for the AHA for disaster management, with specific focus on information and knowledge management. They argued that network-based approach can play a critical role through OODA (Observe, Orient, Decide and act) framework of loop-based decision making. With a comparative analysis of different models of people centric, technology centric and process centric, the paper argued that the Boyd model of OODA helps in making actionable knowledge. The argument was more linked to how the information and knowledge management can be linked to the recovery process as well as its application to other potential hazards, which may occur in the future.

Pollet and Cummins (2009) has argued for AHA for critical infrastructures. They have highlighted that for critical infrastructures, it is a combination of physical security (which may include the risk from natural hazards as well as other conflict related hazards) and cyber security to ensure the smooth operation of the facilities. Apart from that, other functional issues include utility services (like electricity or water supply), effective personnel, situational awareness and emergency management. In total, a number of 54 indicators are provided, which can assess the interlinkages of different types of security through an AHA for critical infrastructures.

Adini et al. (2012), citing examples from public health field have argued the importance of AHA in preparing for mass casualty events (MCE), mass toxicological/chemical events (MTE) and biological events like pandemic. One of the critical suggestion of the analysis was that the building block of Standard Operation Procedure (SOP) was AHA, and its training, emergency drill, preparedness and mitigation aspects. Bayntun (2012) has made a systematic review of disaster management documents and has argued the importance of incorporation of health system approach to all hazard disaster management. This is a direct relation to the Health Emergency Disaster Risk Management (HEDRM) principles and methods, promoted by WHO (World Health Organization). In a similar attempt, Penta et al. (2017) have tried to link the health hazards with natural hazards like earthquake. The arguments are: (1) convergence of people, material, information and resources in case of both public health emergency as well as emergencies created by natural hazards, (2) crisis communication strategies have similar underlying factors, and (3) both have similar demands on medical system. Especially, when it becomes personalized impact at individual level, the link between health and natural hazards become more prominent.

Alertmedia (2022) has defined AHA as:

> The all-hazards approach is a comprehensive emergency preparedness framework that takes the full scope of emergencies or disasters into account when planning for response capacities and mitigation efforts. This means you are prepared for "all hazards" you might face.

It focuses on IT outrages, natural hazards including weather related hazards, power failure, wildfire/bush fire/forest fire, public health related hazards, cybersecurity, conflict, failure of infrastructures etc.

Bodas et al. (2020) have made an interesting analysis of AHA and argued that the "top hazard approach (THA)" possibly is a more feasible option than AHA. This is especially relevant in the resource constraint environment, especially when one needs to prioritize certain hazards over the other to allocate funding and making decisions for investment for preparedness. This is also context specific where the local risks become more critical to local vulnerability, exposure and capacity and therefore certain level of hierarchical decision making to prioritize risks become important. Pelag et al. (2021) have strengthened this argument with COVID-19 pandemic analysis as to why THA is more preferred than AHA. It calls for a different level of planning, response mechanism, mitigation efforts than AHA. The core argument was that COVID-19 did not affect the whole part of the society equally, and therefore the risk mitigation efforts need to be customized based on the intrinsic risks.

In summary, there are two major arguments evolved for all hazards approach over time. One group of researchers argue the importance of AHA by citing public health hazards, cyber security hazards etc., especially for the critical infrastructures as well as individual level hazards. The other group of researchers argue the concept of top hazards, instead of AHA, and the core to that argument is hazards may be common, but the risks are context specific based on the exposures and vulnerability.

18.3 Opportunities of AHA

Centers for Disease Control and Prevention (CDC 2013) of the Office of Public Health Preparedness and Response of the United States has developed a guide for preparedness for AHA, which focuses on natural, chemical, biological, explosion/trauma and radiological hazards. Three steps process are suggested: get a kit, make a plan and be informed. A few kits have been suggested to cope with different types of hazards, formats of plans are provided, which needs customization based on the context and information updating becomes critical since there are dynamic changes in the hazards and risk landscapes. The document also provides examples of different hazards, be it natural or other types of hazards. One of the key messages is to maximize the resources to cope to different types of hazards. It is interesting to note that the key mission and vision of the document is related to protection and security of people's health. The inclusivity approach is also reflected with specific mention on the people with special needs, like physically or mentally challenged people or people with special dietary needs etc.

Kaiser Permanente (2021) has come up with a unique hazard vulnerability assessment (HVA) tool taking into consideration probability and severity. The probability is the likelihood of hazard occurrence, and it deals with sixteen different types of hazards, some of them are natural hazards, some are human induced and conflict related, some are relation to failure of cybersecurity and IT (information technology). Three levels of likelihoods are described: low, medium and high. The other part of the analysis is the severity, which is considered as a function of magnitude and mitigation. The impacts are seen on human, property and business, and the mitigation issues are preparedness, internal response and external response. Based on the analysis, the level of risk is decided as percentage. The unit of the analysis can be a critical infrastructure like school or hospital and government offices. The unit can also be extended to a community or neighborhood. The key issue of the analysis is to focus on different types of potential hazards, where any asset or people or infrastructure are exposed to.

Although the above two documents (both the guidelines and vulnerability and risk assessment tools) provide scope for AHA assessment and preparedness both at the individual level as well as infra or asset unit level (like school or hospital or other critical infra), however, its penetration and usage has been still limited. There are critical issues and challenges which have prohibited the wider usage and implementation of the AHA, especially in assessment or decision making. This part is elaborated in the next section.

18.4 Critical Challenges

Risk assessment: One of the critical challenges is related to risk assessment. A proper risk assessment needs high resolution data related to different types of hazards. During data, there are three core challenges. First, the ownership of the data. In many cases the data belong to different ministries or departments and sharing of data in a common platform often becomes a challenge. For example, the seismic or climate data belongs to the meteorological department. However, when it is health hazards, it belongs to the health department or ministry, and when it becomes technological or radiological hazards, it belongs to the ministry related to industries and trade. While the local government is often in-charge of preparing hazard maps, the lack of data accessibility often urges them to prepare hazard specific maps, not an all hazards map. The second issue is related to openness of data. While there have been different arguments, good practices on open data (Kanbara and Shaw 2022, Kanbara et al. 2022) and its usability, often data is not shared with the argument of being "customized" or "data privacy". In most countries, there is no national open data policy, which distinguishes the nature of data, its accessibility and openness. The third issue is data updating and citizen participation. After the 2018 West Japan flood, Kurashiki city has developed an Open Data platform with citizen participation, which has given citizens an option for updating their experiences from past disasters, like access to evacuation center, evacuation roots etc. This type of regular updating of hazards and risk data with citizen participation not only makes it a robust map, but also engage communities in enhancing their knowledge and perception.

Risk perception: This is also related to some of the issues which are mentioned in the earlier section. Even within the natural hazards, there are challenges of perception between low frequency and high frequency or slow onset or fast onset hazards. For example, people have higher perceptions when it comes to annual events like typhoon or flood, but for earthquake of volcano, the risk perception challenges are higher. Also, for the visible disasters like typhoon or earthquake, the perception may be high, but for slow onset disasters like drought, sea level rise, the challenges are higher. Thus, within natural hazards there are different levels of perceptions. For other invisible hazards like radiological/nuclear or biological or technological, information and trust in information becomes very critical. Transparent information becomes core to the trust in the invisible disaster. It also becomes a moral or ethical obligation from the industry side to provide the right information to the people and communities as well as to the local governments. Possibly, this can be linked to environmental disclosure issues, where there can be clear incentives to the industries to provide right and transparent information. While digital media can and does make a breakthrough for raising the perception with different innovative tools, it also needs to be inclusive, keeping in the mind the needs and priorities of the aging population, who are often not digitally native.

Risk mitigation: Resource prioritization is the key and foremost challenge for risk mitigation. While the risk assessment is done using AHA, the key question is to put appropriate resources to mitigation key risks. There, the top hazards approach becomes important, which may help in prioritizing resources for undertaking mitigation measures. The other challenge is the differential responsibilities in mitigation measures. While there is one department responsible for the natural hazards, different departments are responsible for health or cyber security or technological hazards. Public private people academia partnership becomes core to address some of the challenges related to mitigation. In some cases, it may be easier to allocate mitigation fund for the imminent or regular hazards, however for the low frequency, high consequence event, it needs strong leadership in the decision making.

Governance: Governance is related to decision making, regulation, leadership, funding, managing human resources as well as providing a citizen interface. During the pandemic time of 2020–2022, we have seen strong implementation of adaptive governance, where new measures, adjustments in regulatory framework, innovative decision making, mobilizing and implementing citizen governance interface become critical in many instances. As it is said that urgency is the root of innovation, the urgency during COVID-19 has developed new and innovative governance mechanism. The momentum of innovation should not die because COVID-19 is over, rather it should take this momentum to develop new ways of adaptive governance which can draw significant lessons from the pandemic management and urge for AHA as a new adaptive governance mechanism. Business as usual would not be able to address the critical issues of complex risk landscape.

18.5 Postscripts

Three key points can be mentioned as the postscripts:

AHA as assessment, THA for preparedness: All Hazards Approach (AHA) can be considered as the base for different types of risk assessment, which can identify and prioritize all different types of risks whether at the asset level or individual level or community/city level. Then, we need to adopt the Top Hazard Approach (THA) to put specific planning, awareness and mitigation measures. So, a combination of AHA and THA is required.

Synergy at local level is the key issue: There is strong argument of synergy between development, climate and disaster fields, with the three key global frameworks like Sustainable Development Goals (SDGs), Paris Agreement and Sendai Framework for Disaster Risk Reduction (SFDRR) having similar time horizon, until 2030. For making a resilience and sustainable society and community, it is very important to have a local level implementation, as argued by Shaw et al. (2016). Synergy of these three issues is critical for the AHA, which links and affect different parts of the society differentially.

All hazards, whole government, whole of society: To link all hazards, it is required to link to all different governance mechanisms, as well as the need to address all parts of the society, it is strongly argued that all hazards, whole government and whole society approaches need to be connected closely. Several UN (United Nations) documents have mentioned these separately based on their own needs and priorities, but a real all hazard approach needs to connect all these things together.

References

Adini B, Goldberg A, Cohen R, Laor D, Bar-Dayan Y (2012) Evidence-based support for the all-hazards approach to emergency preparedness. Israel J Health Policy Res 1:40. http://www.ijhpr.org/content/1/1/40

Alertmedia (2022) the all-hazards approach to emergency planning explained. https://www.alertmedia.com/. Accessed on 21 Aug 2023

Bayntun C (2012) A health system approach to all-hazards disaster management: A systematic review. PLoS Curr 4:e50081cad5861d

Bodas M, Kirsch TD, Peleg K (2020) Top hazards approach—rethinking the appropriateness of the all-hazards approach in disaster risk management. Int J Disaster Risk Reduct 47(2020):101559

CDC (2013) All hazards preparedness guide by CDC. https://www.cdc.gov/cpr/documents/ahpg_final_march_2013.pdf. Accessed on 21 Aug 2023

FEMA (1996) Guidance for all hazard emergency operation planning. Federal Emergency Management Agency (FEMA), USA, p 279

Kaiser Permanente (2021) Hazards vulnerability analysis. https://www.calhospitalprepare.org/hazard-vulnerability-analysis. Accessed on 21 Aug 2023

Kanbara S, Shaw R, Kato N, Miyazaki H, Morita A (2022) Society 5.0, digital transformation and disasters: past, present and future. Springer, p 217

Kanbara S, Shaw R (2022) Disaster risk reduction regime in Japan: An analysis in the perspective of open data, open governance. Sustainability 14:19. https://doi.org/10.3390/su14010019

Pelag K, Bodas M, Hertelendy AJ, Kirsch TD (2021) The COVID-19 pandemic challenge to the all-hazards approach for disaster planning. Int J Disaster Risk Reduct 55(2021):102103

Penta S, Marlowe VG, Gill K, Kendra J (2017) Of earthquakes and epidemics: examining the applicability of the all-hazards approach in public health emergencies. Risk Hazards and Crisis in Public Policy https://doi.org/10.1002/rhc3.12112

Pollet J, Cummins J (2009) All hazards approach for assessing readiness of critical infrastructure. IEEE. https://doi.org/10.1109/THS.2009.5168060

Shaw R, Prabhakar SVRK, Chiba Y (2016) Policy brief, SDGs, DRR and CCA: potentials for strengthening interlinkages. IGES, Hayama, Japan, p 12

Vol Lubiz DKJE, Beakley JE, Patricelli F (2008) 'All hazards approach' to disaster management: the role of information and knowledge management. Boyd's OODA Loop, and Network-Centricity, in Disasters 32(4):561–585

WEF (2020) The global risks report 2020. World Economic Forum. https://www.weforum.org/reports/the-global-risks-report-2020/. Accessed on 23 Dec 2022

WEF (2021) The global risks report 2021. World Economic Forum. https://www.weforum.org/reports/the-global-risks-report-2021/. Accessed on 23 August 2023

WEF (2022) The global risks report 2022. World Economic Forum. https://www.weforum.org/reports/the-global-risks-report-2022/. Accessed on 23 Aug 2023

WEF (2023) The global risks report 2023. World Economic Forum. https://www.weforum.org/reports/the-global-risks-report-2023/. Accessed on 23 Aug 2023

Correction to: Adapting All-hazards Planning to Chemical, Biological, Radiological, and Nuclear (CBRN) Threats

R. Alexander Hamilton and John Jones

Correction to:
Chapter 4 in: T. Izumi et al. (eds.), *All-Hazards Approach*, Disaster Risk Reduction, https://doi.org/10.1007/978-981-97-1860-3_4

The original version of the book was inadvertently published with an incorrect Chapter 4 title "Adapting All-hazards Planning to Chemical, Biological, Radioactive, and Nuclear (CBRN) Threats" which has now been corrected as "Adapting All-hazards Planning to Chemical, Biological, Radiological, and Nuclear (CBRN) Threats". The correction chapter and the book has been updated with the changes.

The updated version of this chapter can be found at
https://doi.org/10.1007/978-981-97-1860-3_4

T. Izumi et al. (eds.), *All-Hazards Approach*, Disaster Risk Reduction,
https://doi.org/10.1007/978-981-97-1860-3_19

Correction to: Adapting All-hazards Planning to Chemical, Biological,

FSC
www.fsc.org
MIX
Papier aus verantwortungsvollen Quellen
Paper from responsible sources
FSC® C105338

If you have any concerns about our products,
you can contact us on
ProductSafety@springernature.com

In case Publisher is established outside the EU,
the EU authorized representative is:
Springer Nature Customer Service Center GmbH
Europaplatz 3, 69115 Heidelberg, Germany

Printed by Libri Plureos GmbH
in Hamburg, Germany